RANDOM FIELDS AND SPIN GLASSES

A Field Theory Approach

Disordered magnetic systems enjoy nontrivial properties which are different from and richer than those observed in their pure, non-disordered counterparts. These properties dramatically affect the thermodynamic behaviour and require specific theoretical treatment.

In this book the authors deal with the theory of magnetic systems in the presence of frozen disorder, and in particular paradigmatic and well known spin models such as the Random Field Ising Model and the Ising Spin Glass. They describe some of the most successful approaches to the physics of disordered systems, such as the replica method and Langevin dynamics, together with lesser known results in finite dimension. This is a unified presentation using a field theory language which covers mean field theory, dynamics and perturbation expansion within the same theoretical framework. Particular emphasis is given to the connections between different approaches such as statics vs. dynamics, microscopic vs. phenomenological models. The book introduces some useful and little known techniques in statistical mechanics and field theory including multiple Legendre transforms, supersymmetry, Fourier transforms on a tree, infinitesimal permutations and Ward Takahashi Identities.

This book will be of great interest to graduate students and researchers in statistical physics and basic field theory.

CIRANO DE DOMINICIS is affiliated with the Service de Physique Théorique, Centre d'Etudes de Saclay, France. He has won several prizes including the Langevin prize and Ricard prize from the French Physical Society, the Ampere prize from the French Academy of Sciences, and the Silver Medal of the Centre National de la Recherche Scientifique (CNRS).

IRENE GIARDINA is a researcher at the Istituto Nazionale di Fisica della Materia, Consiglio Nazionale delle Ricerche (CNR), and is also affiliated with the Department of Physics at the University of Rome La Sapienza, Italy.

RANDOM FIELDS AND SPIN GLASSES

A Field Theory Approach

CIRANO DE DOMINICIS
Service de Physique Théorique, Saclay
Commissariat à l'Energie Atomique, France

IRENE GIARDINA
Istituto Nazionale di Fisica della Materia, Roma
Consiglio Nazionale delle Ricerche, Italy

CAMBRIDGE UNIVERSITY PRESS
Cambridge, New York, Melbourne, Madrid, Cape Town, Singapore,
São Paulo, Delhi, Dubai, Tokyo, Mexico City

Cambridge University Press
The Edinburgh Building, Cambridge CB2 8RU, UK

Published in the United States of America by Cambridge University Press, New York

www.cambridge.org
Information on this title: www.cambridge.org/9780521143554

© C. De Dominicis and I. Giardina 2006

This publication is in copyright. Subject to statutory exception
and to the provisions of relevant collective licensing agreements,
no reproduction of any part may take place without the written
permission of Cambridge University Press.

First published 2006
First paperback printing 2010

A catalogue record for this publication is available from the British Library

ISBN 978-0-521-84783-4 Hardback
ISBN 978-0-521-14355-4 Paperback

Cambridge University Press has no responsibility for the persistence or
accuracy of URLs for external or third-party Internet Web sites referred to in
this publication, and does not guarantee that any content on such Web sites is,
or will remain, accurate or appropriate.

To Florence, Ariane,
Marion and Bruno

To Elsa and Andrea

Contents

Preface

I vividly remember the academic year 1977–1978. I was a Loeb lecturer at Harvard that year. The Wilsonian revolution had been blossoming everywhere and I was teaching 'field theory approach to critical phenomena' in the wake of the works of my colleagues and friends Edouard Brezin and Jean Zinn Justin. I had not yet been exposed to the novel intricacies that were being uncovered in the critical behaviour of quenched random systems. But, during that year, several seminars were to deal with them and I began learning and interacting with Mike Stephen, Jo Rudnick and the late Sheng Ma. This is how it all started for me. A good quarter of a century later, the two central problematic systems of the field, the Random Field Ising Model and the Ising Spin Glass, despite several thousand papers and a huge amount of efforts dedicated to them, remain objects of controversy for what concerns how to describe their glassy phase. So why add a book on top of that? Perhaps I will tell how it all occurred.

At the origin the book was a mere set of lecture notes for a course given in this laboratory, a course that was largely repeated two years ago in the theory group of the physics department at UFRS in Porto Alegre. The Lecture Notes Series of Cambridge University Press having been discontinued, it was gracefully suggested that the notes be transformed into a book. I was lucky enough to have had Irene Giardina visiting as a postdoc here. She, as a learned student of Giorgio Parisi, was able to cast a critical ear and eye on the lectures, and then accepted to join in transforming the set of notes into a book. It must be said frankly that, without Irene, the book would not have been born, while with her expertise, several chapters were thoroughly remade and bear her imprint.

In its final version, one third of the book is dedicated to the statics and dynamics of the Random Field Ising Model, one half of it to the Ising Spin Glass, the rest being occupied by the statics and aging dynamics of spherical spins, and by Chapter 5, the lecture delivered by Marc Mézard on the Random Energy Model and the Simplest

Spin Glass. One motivation was to clarify and unify via field theory language what were often found as cryptic and very pointed papers, and to present in some details a few chosen results via useful but little known techniques (e.g. dynamics vs. replicas, multiple Legendre Transforms, supersymmetry vs. Fluctuation–Dissipation Theorem, Fourier Transforms on a tree, infinitesimal permutations vs. Ward–Takahashi Identities). On the controversy that concerns spin glasses, we have chosen to develop the viewpoint according to which the Parisi approach that gives the exact solution for the Sherrington–Kirkpatrick model (i.e. the spin glass in infinite dimension) remains a reasonable starting point to work out what is happening in finite dimension. In contrast with the alternative approach, the so-called droplet theory, which is briefly reviewed in the last chapter, this spin glass field theory approach is a microscopic one in the same sense as the ϕ^4 field theory is a microscopic approach for the pure Ising ferromagnet. A microscopic formulation for a spin glass with droplet-like characteristic properties, that would put both approaches on the same footing, is yet to emerge.

The expertise I may have gathered on the complex subjects presented in this book is, to a considerable extent, the product of enduring discussions I have pursued in the course of years with colleagues and friends. First and foremost I want to mention Edouard Brezin. I remain deeply indebted to him for all the benefits I drew from our discussions, and for the pleasure derived from work accomplished together, some of which is largely used in several chapters of the book. Irene and I are also warmly thankful to Marc Mézard for allowing us to present his views in Chapter 5 and for illuminating exchanges. I am also very grateful to colleagues and friends from the centre, with whom I had inspiring discussions for many years, Alain Billoire, Giulio Biroli, Jean-Philippe Bouchaud, Philippe di Francesco, Thomas Garel, Henri Orland, Jean Zinn-Justin and many others. My thanks also go to my faraway coworkers from past and present with whom I learned so much, Jairo de Almeida from Universidade Federal de Pernambuco in Recife, Andrea Crisanti from Rome University, Imre Kondor from The Collegium, Budapest, Iveta Pimentel from Lisbon University, Tamas Temesvari from Budapest University and Peter Young from California University in Santa Cruz. Loic Bervas typed the lecture notes version on which the book was built and I would like to warmly thank him. Finally I am very grateful to the Service de Physique Théorique and its Director for having extended to us all facilities to conclude this project.

Cirano De Dominicis

My collaboration and friendship with Cirano go back to the period I spent as a postdoc in Saclay, in 1999–2001. The idea and nucleus of this book came out at that time when Cirano gave a set of lectures and we started collaborating on spin glasses. My perspective on the subject was somewhat different from his, being mostly based on the analysis of solvable models and tied to a direct intuitive interpretation of the results. On the other hand, spin glass field theory may sometimes appear obscure, due to the technical complications arising from the nature of the order parameter. I was lucky enough to find someone like Cirano, undisputed master in the field, who introduced me to the field theory approach and who was always ready to discuss any single issue. This book, I hope, bears also the mark of our numerous discussions, of his effort to explain the many subtleties of the theory, and of my attempts to link them with my background and previous works. The writing of the book, was for us the occasion to present in a unified perspective some of the most striking features of disordered systems models. In this light, we added a few more subjects to the original structure of the lecture notes, but decided not to include an explicit treatment of the dynamics of spin glasses. This is surely one of the most promising approaches to these systems, but would have required too much space, rendering the book disproportionate. We have addressed in detail only the simpler case of a disguised ferromagnet, while for real spin glasses we limited ourselves to introducing some of the main concepts characterizing off-equilibrium behaviour and to quoting some important results.

I had the chance to work and discuss on disordered systems with many people, and benefitted from fruitful and stimulating exchanges. There are, however, a few people who had a prominent role in my experience as a physicist and who deserve explicit acknowledgement. I would like to thank first Andrea Cavagna for sharing with me passion, curiosity and enthusiasm in our numerous collaborations. He was kind enough to read extensive parts of this book and his comments have always been precious to me. I am greatly indebted to Giorgio Parisi for having taught me so many things in the past, and for all the inspiring discussions we keep having here in Rome. Cirano and I also thank him warmly for carefully reading this manuscript and for his comments and criticisms. In Oxford, as a postdoc, I worked with David Sherrington and benefitted from his great experience on spin glasses, to him goes my deepest gratitude. I warmly thank Marc Mézard and Olivier Martin for the many discussions we had on spin glasses while I was in Paris, and particularly Jean-Philippe Bouchaud with whom I enjoyed working together on several different subjects. I am also thankful to Alan Bray and Mike Moore for their critical and stimulating comments on my work, during my frequent visits to Manchester in that period. Here in Rome there are many friends and colleagues with whom I constantly discuss on complex systems, and I am grateful to them all. In particular, Enzo Marinari, with whom I also had the pleasure of collaborating on some of the issues discussed in this book, Federico Ricci-Tersenghi and Tomas Grigera, who is now in La Plata. Finally, I would like to thank Alba and Emilio Giardina for their constant support, and for their practical help last summer. I would like to acknowledge the support to this project of the Department of Physics of the University of Rome La Sapienza and of the Institute of Complex Systems of the National Research Council ISC-CNR.

Irene Giardina

Abbreviations

AT	Almeida Thouless
BRST	Becchi–Rouet–Stora–Tyutin
EA	Edwards–Anderson
FDT	Fluctuation–Dissipation Theorem
FT	Fourier Transform
IR	infrared
L-A	longitudinal-anomalous
MSR	Martin–Siggia–Rose
REM	Random Energy Model
RFIM	Random Field Ising Model
RFT	Replica Fourier Transform
RG	Renormalization Group
RS	Replica Symmetric
RSB	Replica Symmetry Broken
SK	Sherrington–Kirkpatrick
TAP	Thouless–Anderson–Palmer
TTI	time translational invariance
UV	ultraviolet

1

A brief introduction

Historically, this book started as a series of lectures given in the Service de Physique Théorique at Saclay, one of the lectures (Chapter 5 here) delivered by Marc Mézard. This explains the strong field theory bias adopted in its approaches and the use of some techniques rarely found in standard literature. It deals with the theory of disordered magnetic systems and for a large part of it with the Random Field Ising Model (RFIM) and the Ising Spin Glass, paradigmatic systems of frozen disorder. Such systems enjoy nontrivial properties, different from and richer than those observed in their pure (nondisordered) counterpart, that dramatically affect the thermodynamic behaviour and require specific theoretical treatment.

Disorder induces frustration and a greater difficulty for the system to find optimal configurations. Consider, for example, the case of spin glasses. These systems are dilute magnetic alloys where the interactions between spins are randomly ferromagnetic or anti-ferromagnetic. They can be modelled using an Ising-like Hamiltonian where the bonds between pairs of spins can be positive or negative at random, and with equal probability. Due to the heterogeneity of the couplings, there are many triples or loops of spin sequences which are *frustrated*, that is for which there is no way of choosing the orientations of the spins without frustrating at least one bond (Toulouse, 1977). As a consequence, even the best possible arrangement of the spins comprises a large proportion of frustrated bonds. More importantly, since there are many configurations with similar degree of frustration, one may expect the existence of many local minima of the free energy.

In mean field models the effects of frustration are enhanced, and several analytical approaches, that are extensively discussed in this book, allow an exhaustive description of the free energy landscape and the thermodynamic behaviour. For spin glasses the scenario that emerges is novel and surprising. The low temperature phase is characterized by ergodicity breaking into an ensemble of hierarchically organized pure states and the static order parameter is a function describing the structure in phase space of such an ensemble (Mézard, Parisi and Virasoro, 1987).

Interestingly, many features predicted by mean field models, such as the behaviour of susceptibilities and correlation functions or the occurrence of aging and off-equilibrium dynamics, are qualitatively observed in experiments, suggesting that the mean field scenario may hold for finite dimensional systems also. To investigate this hypothesis we analyze a field theory for the fluctuations around the mean field solution and discuss its consistency.

A transition to a glassy phase with many pure states seems to occur also in the Random Field Ising Model, where disorder is present as a random external magnetic field contrasting with the ferromagnetic spin–spin interactions. Again, an effective field theory can be developed and the effect of disorder analyzed in detail.

Before addressing in a systematic way some specific disordered models, in this brief introductory chapter we would like to discuss a few general conceptual and technical points that will often recur throughout this book.

1.1 Quenched and annealed averages

In the following we deal with spin models where the disorder is assumed to be *quenched*. What this means is that the 'disordered' variables remain fixed while the spins fluctuate. From an experimental point of view, this corresponds to a situation where the dynamical time scale of the disorder (e.g. the spin couplings in a spin glass) is much longer than the dynamical time scale of the spin fluctuations. In a given experimental sample the disordered variables assume a well defined (though unknown) time independent value. The description in terms of random variables must then be interpreted as follows: each given realization of the random variables corresponds to a given sample of the system, while the distribution according to which they are drawn describes sample to sample fluctuations. A different situation occurs when the disorder is *annealed*, that is when the time scale of the disorder and the one of spin fluctuations are comparable. In this case, in a given experimental sample, the disordered variables vary in time, their statistics being described by the corresponding distribution. The role of time scales in systems with disorder is discussed in Palmer (1982).

For quenched disorder there is a hierarchy between the spin (fast) variables and the disordered (slow) ones which is crucial in many respects. Consider, for example, the thermodynamics. Ideally one would like to compute, for a given sample, averages over the Boltzmann measure and obtain the equilibrium properties of that sample. However, due to the presence of the disorder, one can only compute quantities which are averaged also over the disorder distribution. An important question is thus to understand to what extent these averaged quantities describe the single sample physics.

It turns out that extensive observables, such as the free energy, are particularly well behaved since their associated densities are *self-averaging* in the thermodynamic limit†, that is they assume the same value for each realization of the disorder which has a finite probability. In this case sample to sample fluctuations are vanishing as the volume of the system is sent to infinity and the average value coincides with the *typical* one (i.e. the one assumed in a probable sample). On the contrary, variables that are *not* self-averaging may fluctuate widely from sample to sample and, when computing averages over disorder, rare samples with vanishing probability may give a finite contribution. The self-averageness of extensive variables means that these are the quantities one needs to compute to describe appropriately the behaviour of a single physical system. From a technical point of view, this fact makes many computations more difficult than usual. Let us consider for example the spin glass, where the disorder appears as random couplings (e.g. J_{ij} in the Ising-like model). To describe the thermodynamics of this system we may look at the free energy, which is an extensive variable and is therefore self-averaging. The free energy density f_J for a given disorder realization J is defined as

$$f_J = -\frac{1}{\beta N} \ln Z_J = -\frac{1}{\beta N} \ln \operatorname*{tr}_{\{S_i\}} \exp\{-\beta \mathcal{H}_J\{S_i\}\}, \tag{1.1}$$

where Z_J is the partition function of the model. The average value over the disorder distribution is then given by

$$f = \int \mathrm{d}J\, P(J) f_J = \overline{f_J} = -\frac{1}{\beta N} \overline{\ln Z_J}, \tag{1.2}$$

where we have indicated with an overbar the average over the disorder distribution $P(J)$. In this expression one needs to perform the average of a logarithm, which is not simple to do and quite unusual in statistical mechanics. This is a consequence of the quenched nature of the disorder, which requires us to average extensive observables like the free energy rather than, for example, the partition function itself. For this reason, Eq. (1.2) is usually referred to as a *quenched* average. Note that two distinct averages appear in Eq. (1.2) and in a precise sequence: first the thermodynamic average over the Boltzmann measure which is used to compute f_J, and then the average over the disorder. A much simpler computation is obtained by averaging directly the partition function over the disorder and then taking the logarithm

$$f_{\mathrm{an}} = -\frac{1}{\beta N} \ln \overline{Z_J} = -\frac{1}{\beta N} \ln \int \mathrm{d}J\, P(J) \operatorname*{tr}_{\{S_i\}} \exp\{-\beta \mathcal{H}_J\{S_i\}\}. \tag{1.3}$$

† This can be seen with standard thermodynamic arguments for short range models, and has also been recently proved for long range ones (Guerra and Toninelli, 2002a,b).

In this case the Boltzmann measure and the disorder distribution appear on the same footing and the two corresponding averages are performed at the same time. This procedure would be appropriate were the disorder of an annealed kind, for this reason Eq. (1.3) is referred to as an *annealed* average.

1.2 The replica method

An indirect way to deal with the logarithm appearing in the quenched average Eq. (1.2) relies on the so-called *replica method* (Kac, 1968; Edwards, 1972). This method is based on the following elementary relationship:

$$\ln Z = \lim_{n\to 0} \frac{Z^n - 1}{n}. \tag{1.4}$$

Thanks to Eq. (1.4) the average of the logarithm is reduced to the average of Z^n. For integer n this can be expressed as the product of the partition functions of n identical copies, or *replicas*, of the original system. In this way, we have

$$\overline{\ln Z} = \lim_{n\to 0} \frac{\ln \overline{Z^n}}{n} = \lim_{n\to 0} \frac{1}{n} \ln \overline{\operatorname*{tr}_{\{S_i^a\}} \exp\left\{-\beta \sum_a \mathcal{H}_J{}^a\{S_i^a\}\right\}}, \tag{1.5}$$

where a is a replica index. The average over the disorder appearing in the r.h.s. of (1.5) is now of a standard kind and can be carried out with simple algebra. Likewise, if one wishes to average over disorder an observable like a correlation function, e.g.

$$C_{jk} = \frac{1}{Z_J} \operatorname*{tr}_{\{S_i\}} \exp\{-\beta \mathcal{H}_J\{S_i\}\}\, S_j S_k, \tag{1.6}$$

one needs to resort to replicas again to get rid of the J dependence of the norm. Multiplying top and bottom by Z_J^{n-1} gives

$$\begin{aligned} C_{jk} &= \frac{Z_J^{n-1}}{Z_J^n} \operatorname*{tr}_{\{S_i\}} \exp\{-\beta \mathcal{H}_J\{S_i\}\}\, S_j S_k \\ &= Z_J^{-n} \operatorname*{tr}_{\{S_i^a\}} \exp\left\{-\beta \sum_a \mathcal{H}_J{}^a\{S_i^a\}\right\} S_j^1 S_k^1, \end{aligned} \tag{1.7}$$

and finally

$$\overline{C_{jk}} = \lim_{n\to 0} \overline{\operatorname*{tr}_{\{S_i^a\}} \exp\left\{-\beta \sum_a \mathcal{H}_J{}^a\{S_i^a\}\right\} S_j^1 S_k^1}. \tag{1.8}$$

Under disorder averaging, disorder with independent replicas is replaced by *coupled* replicas. The task is then to compute properties of the system with the effective fields and couplings resulting from J-averaging (ϕ_i^{ab} for spin glasses, related to the spin overlap $S_i^a S_i^b$ and ϕ_i^a for the Random Field Ising Model, related to S_i^a) and analytically continue the result to $n = 0$.

In spin glasses, as we shall see, even the mean field approximation is highly nontrivial, due to the matrix nature of the order parameter Q_i^{ab}, the thermal average of the field ϕ_i^{ab}. It turns out that the mean field solution may break the invariance with respect to replica permutations, endowed by the original replicated Hamiltonian. In this case we say that Replica Symmetry Breaking (RSB) occurs. For RSB solutions, deciding which is the correct pattern of symmetry breaking, i.e. what is the structure of the overlap matrix Q^{ab} in the replica space, is a demanding task. We will discuss in detail the correct ansatz for Q^{ab} and the novel physical scenario it describes. This complex RSB structure makes the analysis of the Gaussian fluctuations around the mean field solution much more complicated than in standard systems, and new techniques must be introduced to deal with the inversion of the Hessian matrix.

In the Random Field Ising Model, on the other hand, the mean field solution is trivial since the order parameter bears only one replica index. In this case the treatment of the model in finite dimension is simpler and a perturbative renormalization group can easily be performed. The glassy phase is, however, more difficult to detect, requiring a more sophisticated analysis of the dependence of the free energy on two-point functions.

1.3 The generating functional

So far, we have discussed the conceptual and technical problems originated by the presence of the quenched disorder in static computations. The same kind of difficulties arise when adopting a dynamical approach. Let us consider, for example, a Langevin kind of dynamics, which is the one mostly used in analytical computations. In this case the dynamical evolution of a given field $\phi_i(t)$ is determined by the following stochastic equation:

$$\mathcal{E}_i^J \{\phi_i(t)\} \equiv \frac{\partial \phi_i(t)}{\partial t} + \frac{\partial \mathcal{H}_J \{\phi_i\}}{\partial \phi_i} - \eta_i(t) = 0, \tag{1.9}$$

where $\eta_i(t)$ represents a Gaussian thermal noise. Here, again, two noises appear (the thermal noise and the quenched disorder) and two averages must be performed. In principle, one should first compute, for a given disorder instance, the dynamical observables by integrating out the thermal noise. Then, the result must be averaged over the distribution of the disorder. The *generating functional* approach (Martin, Siggia and Rose, 1973) is a technique which allows us to do it all, in a way that bears resemblance to the static computations. The main idea is to introduce a dynamical functional using the identity

$$1 = \hat{Z} = \int \prod_i \mathrm{D}\phi_i(t)\, \delta\left(\mathcal{E}_i^J\{\phi_i(t)\}\right) \left|\det \partial_j \mathcal{E}_i^J\right|, \tag{1.10}$$

where $\mathrm{D}\phi_i(t)$ stands for functional integration over the field $\phi_i(t)$.

In the presence of appropriate external sources $\hat{Z}$ becomes then the generating functional of dynamical averages (see Chapter 3). Both the delta function and the determinant are expressed using integral representations and the generating functional is written in terms of a dynamical (disorder dependent) Lagrangian which plays for the dynamics a role analogous to the replicated Hamiltonian in the statics (summations over replicas being replaced by integrals over time). At this point, because the norm $\hat{Z}$ is J independent (in contrast to Z_J for the static case) one can trivially perform the average over quenched disorder and over thermal noise. The result is the effective generating functional for correlations and responses (the analogues of the static overlap matrix). As we have seen, in the static computation of free energy, the average over disorder generates a coupling between distinct replicas. In the dynamical context there are no replicas, and the effect of the disorder is to generate nonlocality in time, i.e. a coupling between distinct *times*. In statics, the order parameter may break the replica permutation symmetry and exhibit a nontrivial structure in replica space. In dynamics, correlation and response functions may in some regimes break the time translation invariance and exhibit nonstandard patterns of dynamical evolution where the fluctuation dissipation theorem is violated. We shall discuss such a scenario in detail for a simple spin glass model.

1.4 General comments

The difficulties related to the analysis of disordered models stem mainly from the complex nature of the order parameter and the existence of a glassy phase. In the context of the replica method, this is already manifest at mean field level, where, below the transition to the glassy phase, the saddle point acquires an RSB structure. From a dynamical point of view, time translational invariance is lost and equilibrium never reached. The analysis in finite dimension becomes rather complicated, since even the computation of Gaussian fluctuations around an RSB mean field solution is not a simple problem. In this book, we deal with two classics of disordered models: spin models with disordered magnetic field (the Random Field Ising Model) and spin models with disordered exchange couplings (spin glasses). For these two cases, the above difficulties affect our analysis in different respects:

(i) For the Random Field Ising Model, we are mostly interested in understanding the nature of the transition between the ferromagnetic and the paramagnetic state and computing the critical point properties. To do that, we approach the transition from above, always remaining in a replica symmetric region. (Correspondingly, the dynamics is of an equilibrium, time translational invariant, kind.) In this phase a renormalization group analysis, both static and dynamic, up to one loop can be carried out. The presence of bound states can be investigated exactly at the ferro–para transition, when and if the theory is still replica symmetric. However, the vitreous phase is not directly addressed.

(ii) For spin glasses, a mean field analysis reveals a rich low temperature phase which can be described in detail. Our main aim is then to study the stability of the mean field scenario in finite dimension. To do that, we place ourselves in the low temperature region and develop a field theory for the fluctuations around the mean field RSB solution. Since we now deal with a replica symmetry broken theory, we mainly analyze the Gaussian fluctuations and obtain the free propagators. One-loop corrections in the glassy phase are only dealt with for the equation of state, relying upon scaling arguments to hint at behaviour away from the upper critical dimension. Renormalization group calculations are carried out at the critical temperature.

References

Edwards S. F. (1972). *Proceedings of the Third International Conference on Amorphous Materials, 1970*, ed. R. W. Douglass and B. Ellis, New York, Wiley.
Guerra F. and Toninelli F. L. (2002a). *J. Math. Phys.*, **43**, 6224.
Guerra F. and Toninelli F. L. (2002b). *Comm. Math. Phys.*, **230**, 71.
Kac M. (1968). Trondheim Theoretical Physics Seminar, *Nordita Publ. 286.*
Martin P. C., Siggia E. and Rose H. (1973). *Phys. Rev. A*, **8**, 423.
Mézard M., Parisi G. and Virasoro M. A. (1987). *Spin Glass Theory and Beyond*, Singapore, World Scientific.
Palmer R. G. (1982). *Advances in Physics*, **31**, 669.
Toulouse G. (1977). *Comm. Phys.*, **2**, 115.

2
The Random Field Ising Model

The Random Field Ising Model (RFIM) represents one of the simplest models of cooperative behaviour with quenched disorder, and it is, in a way, complementary to the Ising Spin Glass which will be extensively treated later in this book. It accounts for the presence of a random external magnetic field which antagonizes the ordering induced by the ferromagnetic spin–spin interactions. From an experimental point of view, on the other hand, as shown by Fishman and Aharony (1979) and Cardy (1984), it is equivalent to a dilute anti-ferromagnet in a *uniform* field (see Belanger, 1998 for a recent review on experimental results).

Despite twenty-five years of active and continuous research the RFIM is not yet completely understood. The problem seems related to the presence of bound states in the ferromagnetic phase, which make the standard theoretical approaches not adequate to analyze the critical behaviour. Here we discuss the RFIM in the context of perturbative field theory. The chapter is organized as follows: in Section 2.1 we define the model and outline the main expectations for its qualitative behaviour. In Section 2.2 we introduce an effective replicated ϕ^4 field model where the disorder has been integrated out. Then we perform a perturbative analysis on this model (Section 2.3) and illustrate how the so-called *dimensional reduction* arises (Section 2.4). Finally, in Section 2.5 we introduce some generalized couplings which need to be taken into account to properly describe the system; we perform a perturbative Renormalization Group (RG) close to the upper critical dimension (Sections 2.6 and 2.7) and discuss the occurrence of a vitrous transition (Section 2.8).

We leave aside several other approaches used to treat this model such as real space RG, high temperature expansions, Monte Carlo simulations, etc. For those we refer the reader to the review of Natterman (1998). We should also mention some recent and interesting work by Tarjus and Tissier (2003) that uses the functional RG in a very promising way.

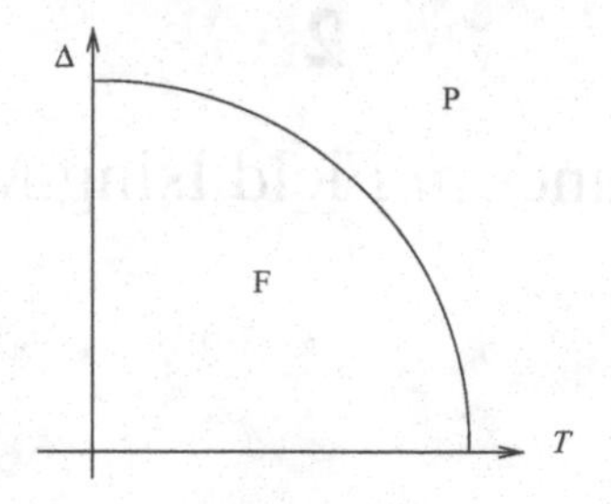

Figure 2.1 Qualitative phase diagram for the RFIM with zero external homogeneous field: P indicates the paramagnetic phase, F the ferromagnetic one

2.1 The model

The Hamiltonian of the RFIM is analogous to the one of the classical Ising Model, but allowing for a disordered quenched magnetic field:

$$-\mathcal{H} = \sum_{(ij)} J_{ij}\, S_i S_j + \sum_i h_i\, S_i . \tag{2.1}$$

Here $S_i = \pm 1$, $J_{ij} = J$ for nearest neighbour pairs (i, j) and the h_i are quenched random variables drawn with a Gaussian distribution defined by

$$\overline{h_i} = 0, \qquad \overline{h_i h_j} = \Delta\, \delta_{i;j} . \tag{2.2}$$

Note that, because of the presence of the quenched disorder, we deal from now on with two different kinds of average: the thermal average over the Boltzmann measure and the quenched average over the disorder distribution. To distinguish them we shall indicate the first with brackets $\langle \cdots \rangle$ and the second with an overbar $\overline{\cdots}$ (as in (2.2)).

In the pure case, i.e. for the Ising Model with no external magnetic field, a second order transition exists at temperature T_c^0, separating a high temperature paramagnetic phase from a low temperature ferromagnetic one. The presence of a random external magnetic field clearly disturbs the ordering effect associated with the ferromagnetic exchange interactions: thus one expects a decrease of the transition temperature with increasing disorder strength Δ. Qualitatively, then the phase diagram exhibits a paramagnetic phase for large Δ, and/or large temperature T, and a ferromagnetic phase in the opposite limits (see Fig. 2.1).

At low enough dimensions the action of the random field can inhibit the creation of the ordered phase. A quite robust argument has been given by Imry and Ma (1975). It estimates how a random field can destroy a predominantly ferromagnetic environment. Consider a domain of size R in a ferromagnetic region (see Fig. 2.2) and reverse the spins inside it. The energy cost due to the exchange interactions E_J is proportional to the surface of the domain and is therefore of order $J R^{D-1}$,

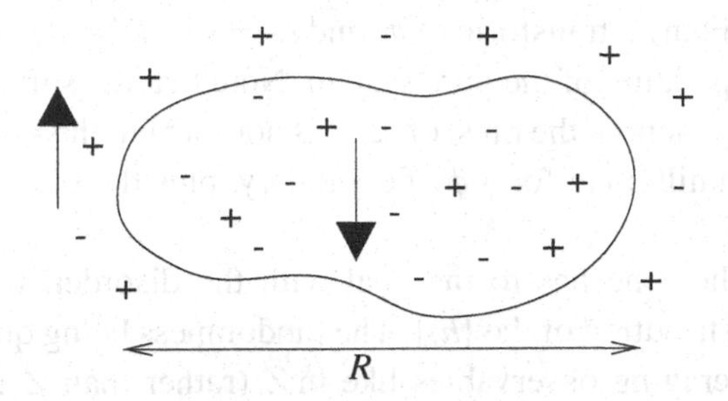

Figure 2.2 Domain of size R in a ferromagnetic environment

where D is the dimension of the physical system. The Zeeman energy associated with the random field E_{RF} is, according to the central limit theorem, $E_{\mathrm{RF}}^2 \sim \Delta R^D$. The global energy balance is then written

$$E(R) \approx J R^{D-1} - \sqrt{R^D \Delta}, \tag{2.3}$$

and the fluctuations of h_i will always destroy the ferromagnetic state if

$$\frac{D}{2} > D - 1, \quad \text{i.e.} \quad D < 2. \tag{2.4}$$

2.2 The replicated field theory

It is convenient to recast the Hamiltonian (2.1) into a soft spin version. This can easily be done by writing, within a constant, the partition function Z as:

$$Z = \int \prod_i (\mathrm{D}\phi_i) \operatorname*{tr}_{\{S_i\}} \left\{ \exp\left[-\frac{1}{2\beta} \sum_{(ij)} \phi_i (J^{-1})_{ij} \phi_j + \sum_i (\phi_i + \beta h_i) S_i \right] \right\}, \tag{2.5}$$

where $\mathrm{D}\phi_i$ stands for $\mathrm{d}\phi_i/\sqrt{2\pi}$. By taking the trace over the spins, we then get

$$Z = \int \prod_i (\mathrm{D}\phi_i) \exp\left\{ -\frac{1}{2\beta} \sum_{i,j} \phi_i (J^{-1})_{ij}\, \phi_j + \sum_i \ln\cosh\,(\phi_i + \beta h_i) \right\}. \tag{2.6}$$

If we assume that the argument of the cosh is small, develop it and retain the relevant terms, after appropriate rescaling, we can in turn rewrite (2.6) as a ϕ_i^4 Lagrangian in a random field h_i:

$$\mathrm{e}^{-F\{h_i\}} \equiv Z = \int \prod_i (\mathrm{D}\phi_i)\ \exp\left\{ -\frac{1}{2} \int \frac{\mathrm{d}p^D}{(2\pi)^D}\, \phi(p)[p^2 + r_0]\, \phi(-p) \right.$$
$$\left. - u_1 \sum_i \frac{\phi_i^4}{4!} + \sum_i h_i\, \phi_i \right\}, \tag{2.7}$$

where[†] $\phi(p)$ is the Fourier transform of ϕ_i and $r_0 = (T/T_c^0 - 1)$, $T_c^0 = J$ being the bare transition temperature of the pure system. Note that the soft spin Hamiltonian appearing in the exponent of the r.h.s. of (2.7) is nothing but the standard Ginzburg–Landau–Wilson Hamiltonian for a ϕ^4 field theory, plus the term with the random field.

To proceed further one has to the deal with the disorder, that is one has to average over the distribution of the $\{h_i\}$. The randomness being quenched, we have to consider self-averaging observables like $\ln Z$ (rather than Z itself). We resort then to the replica trick,

$$\ln Z = \lim_{n\to 0} \frac{Z^n - 1}{n}. \tag{2.8}$$

Z^n can be expressed as the product of the partition function of n identical independent copies (or *replicas*) of the original system, and we have

$$Z^n = \prod_{a=1}^{n} Z_a = \int \prod_{i,a} (\mathrm{D}\phi_i^a)\, \mathrm{e}^{-\beta \sum_a \mathcal{H}\{\phi^a\}}. \tag{2.9}$$

Now we can easily perform the Gaussian average over the random magnetic field and we get:

$$\begin{aligned}\overline{Z^n} &= \int \prod_{i,a} (\mathrm{D}\phi_i^a)\, \exp\left\{ -\frac{1}{2}\sum_a \int \frac{\mathrm{d}p^D}{(2\pi)^D} \phi^a(p)\, [p^2 + r_0]\, \phi^a(-p) \right.\\ &\quad \left. - \frac{u_1}{4!}\sum_{a,i} (\phi_i^a)^4 + \frac{1}{2}\Delta \sum_i \left(\sum_a \phi_i^a\right)\left(\sum_b \phi_i^b\right) \right\}\\ &= \int \prod_{i,a} (\mathrm{D}\phi_i^a) \mathrm{e}^{\mathcal{L}\{\phi_i^a\}},\end{aligned} \tag{2.10}$$

where the upper (replica) indices a, b run from 1 to n. In the exponent of (2.10) we read the effective Lagrangian $\mathcal{L}\{\phi_i^a\}$ with which we deal now. Note that, due to the disorder average, the n replicas which were originally independent become coupled through the Δ term, giving rise to a nontrivial effective field theory. Note also that taking the average with the measure defined in (2.10) is equivalent to taking the average over the equilibrium distribution *and* over the quenched disorder, i.e. $\langle\cdots\rangle_{\mathcal{L}} \equiv \overline{\langle\cdots\rangle}$ (but we shall drop the index $\mathcal{L}$ from the averages when it does not generate confusion).

[†] To lighten the notation, we shall use roman letters also to represent vectors, unless we wish to emphasize the vectorial nature of the corresponding variable, as for example in Chapter 10.

In the next section we proceed with a standard perturbative expansion starting from Expression (2.10).

2.3 Perturbation expansion

To keep things simple, we are interested in what is happening as we go down in temperature (say, at constant Δ) from the paramagnetic phase, down to the Curie line $T_c(\Delta)$. Let us proceed to analyze the effective field theory we are left with.

2.3.1 *Bare propagators*

In (2.10) the quadratic part gives us the matrix

$$\left[G^0\right]^{-1}_{ab}(p) = (p^2 + r_0)\delta_{a;b} - \Delta \tag{2.11}$$

whose inverse is our bare propagator G^0_{ab}. To invert this matrix it is convenient to express it in terms of appropriate projector operators:

$$\left[G^0\right]^{-1}_{ab}(p) = (p^2 + r_0)\left(\delta_{a;b} - \frac{1}{n}\right) + (p^2 + r_0 - n\Delta)\frac{1}{n}. \tag{2.12}$$

Indeed the *operators*

$$(P_T)_{ab} = \delta_{a;b} - \frac{1}{n}, \qquad (P_L)_{ab} = \frac{1}{n}$$

satisfy $P_T^2 = P_T$, $P_L^2 = P_L$, $P_T P_L = P_L P_T = 0$ and are therefore projectors. Hence we get

$$\begin{aligned} G^0_{ab}(p) &= \frac{1}{p^2 + r_0}\left(\delta_{a;b} - \frac{1}{n}\right) + \frac{1}{p^2 + r_0 - n\Delta}\left(\frac{1}{n}\right) \\ &= \frac{\delta_{a;b}}{p^2 + r_0} + \frac{\Delta}{(p^2 + r_0)(p^2 + r_0 - n\Delta)}; \end{aligned} \tag{2.13}$$

graphically

$$\underset{a}{\longrightarrow} + \underset{a\qquad b}{\longrightarrow\!\!\times\!\!\longrightarrow}. \tag{2.14}$$

Note that the first term is the *bare* contribution to the *connected* propagator i.e. $\langle\phi\phi\rangle_c = \langle\phi\phi\rangle - \langle\phi\rangle\langle\phi\rangle$, that is the contribution present even in absence of the random field, which, at the bare level, is unchanged under h-averaging. The second term is the (bare) contribution to the *disconnected* propagator $\langle\phi\rangle\langle\phi\rangle$, which under h-averaging becomes connected.

The physical interpretation is straightforward. Consider two identical copies of the system indexed by a and b. In absence of the field in each copy the magnetization ϕ_i is correlated in space as expressed by the standard bare propagator and fluctuates independently of the other copy, thus $\langle \phi_i^a \phi_j^b \rangle$ is zero for $a \neq b$. Besides, the average local magnetizations $\langle \phi_i \rangle$ are zero for each of the two copies. Therefore $G^0_{ab}(p) = \delta_{a;b}/(p^2 + r_0)$. When we consider the system in presence of the field, the same is true for what concerns the correlations, but not for the average local magnetization. Indeed, given h_i, the local magnetization of each copy $\langle \phi_i^a \rangle$ is nonzero and correlated with the field (even if $\sum_i \langle \phi_i \rangle = M = 0$). Since distinct copies feel the *same* field, whatever its random realization, their magnetizations will then be on average correlated and, at the bare level, $\overline{\langle \phi_i^a \rangle \langle \phi_i^b \rangle} \propto \Delta$.

2.3.2 Vertices

Here we only have one vertex:

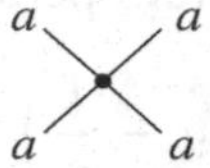

$a \quad a$

$: u_1$

$a \quad a$ (2.15)

To generate all the needed graphs it may be convenient to introduce a *source* term $\sum_{i,a} H_i^a \phi_i^a$ (with H_i different from the random field h_i which has already been integrated out). In this way the effective Lagrangian appearing in (2.10) becomes

$$\mathcal{L}\{\phi_i^a\} = -\frac{1}{2}\sum_{a,b}\int \frac{\mathrm{d}p^D}{(2\pi)^D}\, \phi^a(p)\left[G^0\right]^{-1}_{ab}\phi^b(-p) - \frac{1}{4!}u_1\sum_i \left(\phi_i^a\right)^4 + \sum_{i,a} H_i^a \phi_i^a . \tag{2.16}$$

In the end the source term is set to zero (and then, in the paramagnetic phase, $\langle \phi^a \rangle = 0$).

2.3.3 Perturbation expansion: the free energy

When performing the h-average over Z^n we have

$$\overline{\mathrm{e}^{-F\{h_i\}}} \equiv \overline{Z^n\{h_i\}} = \exp\left\{-n\,\overline{F\{h_i\}} + \frac{n^2}{2}\overline{\left(F\{h_i\} - \overline{F\{h_i\}}\right)^2} + \cdots\right\}. \tag{2.17}$$

The average free energy is the *single replica* contribution. And it is thus exactly the same expansion as for a pure system, except that the bare propagators are given by (2.13) rather than just by the $\delta_{a;b}$ piece of it.

For the *pure* (i.e. no random field) system (in zero H field, and in the paramagnetic phase) we get†

$$-F(H=0) = -\frac{1}{2}\mathrm{tr}\ln[G^0]^{-1} + \cdots + \cdots + \cdots + \cdots, \qquad \frac{1}{2^3} \quad \frac{1}{2}\cdot\frac{1}{4!} \quad \frac{1}{2}\cdot\frac{1}{2^3} \tag{2.18}$$

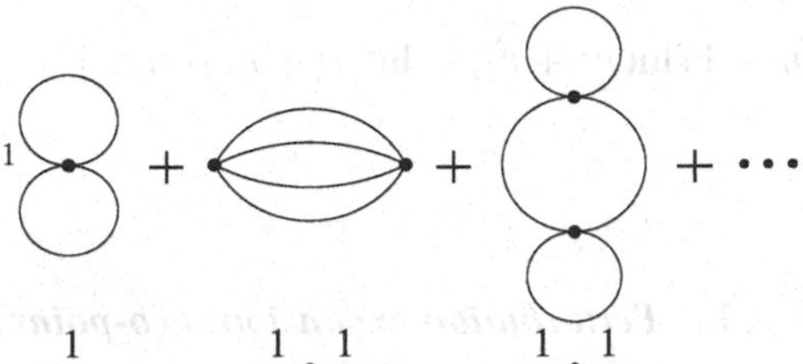

where we have displayed the symmetry weights corresponding to each graph: $(1/2^3)$, $(1/2)\cdot(1/4!)$, $(1/2)\cdot(1/2^3)$.

Proceeding in the same way, for the RFIM we get

$$-n\overline{F\{h_i; H=0\}} = -\frac{1}{2}\mathrm{tr}\ \ln\left[G^0\right]^{-1}_{ab}$$

$$+\sum_a \left\{ \cdots + \cdots + \cdots + \cdots + \cdots + \cdots + \cdots \right\}, \qquad \frac{1}{2^3} \quad \frac{1}{2^2} \quad \frac{1}{2^3} \quad \frac{1}{2}\cdot\frac{1}{4!} \quad \frac{1}{2}\cdot\frac{1}{3!}$$

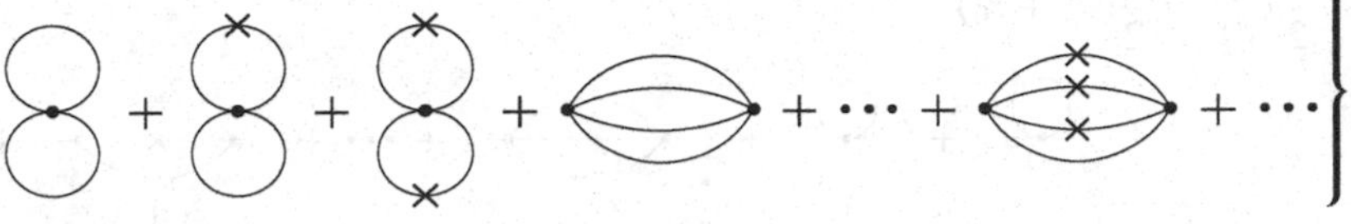

where all the lines bear the identical replica index, a.

Note that terms like

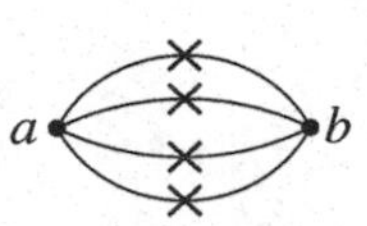

$$a \quad b \qquad \frac{1}{4!} \tag{2.19}$$

bear two free indices a, b, are then of order n^2 and must *not* be included. This graph was disconnected into two pieces before h-averaging and contributes to the term in $(n^2/2)(\overline{F\{h_i\}} - \overline{F\{h_i\}})^2$. Note that all the graphs which contain only crossed Δ propagators necessarily bear at least two indices and must therefore be disregarded at this level.

† On the graphs we read off the $\frac{1}{s}$ factor; s is the symmetry factor ('order of the group of automorphisms of the graph')!

Let us evaluate the term with tr $\ln[G^0]$. Using the projectors, we have (to be integrated over p):

$$-\frac{1}{2}\{(n-1)\ln(p^2+r_0)+\ln(p^2+r_0-n\Delta)\} = -\frac{n}{2}\ln(p^2+r_0) - \frac{n\Delta}{p^2+r_0} + \cdots \tag{2.20}$$

2.3.4 Perturbation expansion: two-point and four-point functions

We are interested in the *critical* region where fluctuations take over. Renormalization theory is the tool that allows us to discuss the behaviour of the theory in that region, and for that we need to exhibit the behaviour of the two- and four-point vertex functions (higher powers of ϕ than ϕ^4 are irrelevant operators).

- *two-point function*
  We have

$$\frac{1}{2}\Gamma^{(2)}_{aa} = \frac{1}{2}(p^2+r_0) \tag{2.21}$$

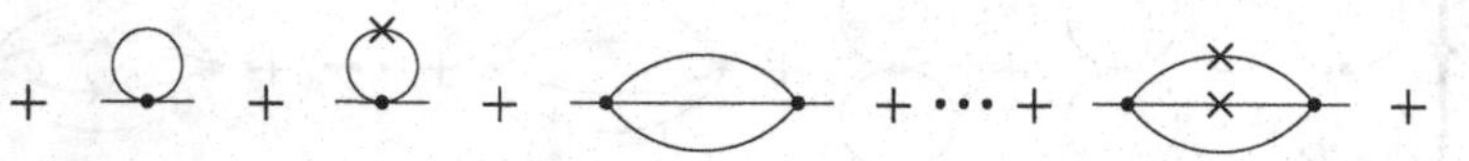

$$+ \quad + \quad + \quad + \cdots + \quad + \cdots,$$

$$\frac{1}{4} \qquad \frac{1}{4} \qquad \frac{1}{2}\cdot\frac{1}{3!} \qquad \frac{1}{2^3}$$

  where we did not include the subleading graph .

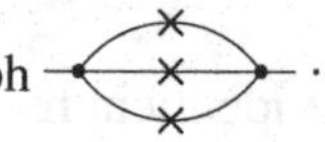

$$\frac{1}{2}\cdot\frac{1}{3!}$$

- *four-point function*
  We have

$$\frac{1}{4!}\Gamma^{(4)}_{aaaa} = \quad + \quad + \quad + 2\text{ loops}, \tag{2.22}$$

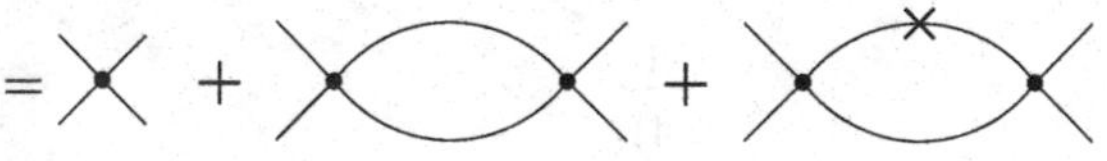

$$\frac{1}{4!} \qquad \frac{1}{2^4} \qquad \frac{1}{2^3}$$

  where we did not include the subleading graph .

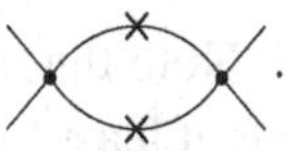

$$\frac{1}{2^4}$$

Here we have exhibited single-replica graphs. In the corresponding contributions to the pure system, the graphs involving *disconnected* propagators would be absent.

2.4 Most divergent graphs and dimensional reduction

In high dimension, mean field is a good description of the system. As the dimension decreases, fluctuations increase. They start dominating the behaviour of the system at the upper critical dimension D_c. To obtain D_c one has to look at the dimension where ultraviolet (UV) divergences develop in the perturbation expansion of the vertex function. Let us first look at the vertex functions for the pure system. Let us count the degree of UV divergence of terms containing V vertices, I internal lines and contributing, for example, to $\Gamma^{(4)}$. If we count the number of independent moments we get a relationship for the number of loops L (momentum conservation occurs in every vertex, and in the whole graph):

$$L = I - (V - 1). \tag{2.23}$$

We also have

$$E + 2I = 4V, \tag{2.24}$$

where $E = 4$ is the number of external lines. Hence

$$L = I/2. \tag{2.25}$$

Then the dimension in p of a graph of L loops is

$$\frac{(p^D)^L}{(p^2)^{2L}} = p^{L(D-4)}, \tag{2.26}$$

from which we see that for $D = 4$ all graphs give a logarithmic divergent contribution. Hence the upper critical dimension for the *pure* system is $D_c = 4$.

Let us now consider the RFIM and evaluate in the same way the upper critical dimension. In this case the global bare propagator differs from the one of the pure system by the off diagonal part, which goes as Δ/p^4 near $T_c(\Delta)$. Thus many graphs will have a contribution from this part of the propagator (the crossed links in Expression (2.22)). The more Δs, the more divergent will be the graph. We have to look at the most divergent but connected (single replica) contributions. Clearly, the best we can do is to take, *before averaging*, the graphs that are richest in h_is, i.e. the tree-graphs. Then under h-averaging, the h_is coalesce in pairs, $\overline{h_i h_j} = \Delta\delta_{i;j}$, and we'll thus have one Δ per loop! That selection of all most divergent graphs contributing to $\Gamma^{(4)}_{aaaa}$ gives a UV divergence like

$$\frac{(p^D)^L}{(p^2)^{2L}(p^2)^L} = p^{L(D-6)} = p^{L((D-2)-4)}. \tag{2.27}$$

Thus the situation is identical to that of the pure system in dimension $D - 2$. This is the meaning of the so-called *dimensional reduction*. This is indeed true to all orders in the perturbation expansion, as shown by Aharony *et al.* (1976) and by Young (1976), who checked the behaviour of multi-loop integrals.

2.4.1 Dimensional reduction and supersymmetry

A complete and compact proof of dimensional reduction due to Parisi and Sourlas (1979) can be obtained using the following remarks.

(i) The tree diagrams of the field theory (2.7) are generated by the saddle point equation

$$0 = \frac{\partial \mathcal{H}\{\phi_i\}}{\partial \phi_i} - h_i, \tag{2.28}$$

where $\mathcal{H}\{\phi_i\} - \sum_i h_i \phi_i$ represents the soft spin Hamiltonian appearing in the exponent of (2.7). This can be checked diagrammatically by expanding (2.28) in powers of h.

A formal way to compute the N-point functions consists of introducing the generating functional

$$\Xi\{\ell_i\} = \int \prod_i \mathrm{D}\phi_i \, \mathrm{e}^{\ell_i \phi_i} \delta\left(\frac{\partial \mathcal{H}}{\partial \phi_i} - h_i\right) \det \mathcal{J}, \tag{2.29}$$

where $\mathcal{J}$ is the Jacobian $\mathcal{J} = (\partial^2 \mathcal{H}/\partial \phi_i \, \partial \phi_j)$. Correlation functions can then be obtained by taking derivatives of (2.29) with respect to the source ℓ_i. Note that $\Xi(0) = 1$, by virtue of the normalized δ functions.

Equivalently, using standard representation of δ functions and determinants, we have

$$\Xi\{\ell_i\} = \int \prod_i (\mathrm{D}\phi_i \, \mathrm{D}\hat{\phi}_i \, \mathrm{D}c_i \, \mathrm{D}\overline{c}_i) \exp\left\{ \sum_i \ell_i \phi_i + \mathrm{i} \sum_i \hat{\phi}_i \left(\frac{\partial \mathcal{H}}{\partial \phi_i} - h_i\right) + \sum_{i,j} c_i \frac{\partial^2 \mathcal{H}}{\partial \phi_i \, \partial \phi_j} \overline{c}_j \right\}, \tag{2.30}$$

where $\hat{\phi}$ is the Lagrange multiplier used to implement the δ function and $(c, \overline{c})$ are the fermion fields used to express the Jacobian.

(ii) Ξ being an observable, we may now trivially do the h-average (no troublesome normalization here since $\Xi(0) = 1$). The effective averaged Lagrangian is

$$\mathcal{L} = \sum_i \left(\hat{\phi}_i \frac{\partial \mathcal{H}}{\partial \phi_i} + \frac{\Delta}{2} \hat{\phi}_i^2\right) + \sum_{i,j} c_i \frac{\partial^2 \mathcal{H}}{\partial \phi_i \, \partial \phi_j} \overline{c}_j. \tag{2.31}$$

(iii) It is then a general property of such systems constrained by a classical equation of motion with a noise (Zinn-Justin, 1989, Chapter 29) to enjoy a supersymmetric invariant formulation (with space x, extended to Grassmann components $\theta, \bar{\theta}$) and with the Lagrangian becoming

$$\int \mathrm{d}\overset{D}{x} \, \mathrm{d}\theta \, \mathrm{d}\bar{\theta} \left\{ \Phi\left(\Delta + \tfrac{\partial}{\partial \theta} \tfrac{\partial}{\partial \bar{\theta}}\right) \Phi + V(\Phi) \right\},$$
$$\Phi = \phi(x) + c(x)\bar{\theta} + \overline{c}(x)\theta + \theta\bar{\theta}\hat{\phi}(x). \tag{2.32}$$

Here θ is such that $\theta^2 = \bar{\theta}^2 = \theta\bar{\theta} + \bar{\theta}\theta = 0$, and V indicates the potential part of the original Hamiltonian. Thus, we get a field theory which is analogous to the one we started with, but living in the supersymmetric space of coordinates $(x, \theta, \bar{\theta})$. Intuitively, with $\boldsymbol{\Delta}$ a Laplacian in D components, to get the dimension of this supersymmetric space, one is to add the dimension d of the space associated with the Grassman variables. But such variables may be attributed a dimension $d = -2$ since $\det \equiv (1/\sqrt{\det})^{-2}$, we then have $D \to D - 2$! More rigorously, one can prove that, due to the supersymmetry endowed by the Lagrangian (2.32), integrals in the supersymmetric space are equivalent to integrals in a $D - 2$ space dimensions (Parisi and Sourlas, 1979).

Dimensional reduction, however, does *not* hold for the RFIM. It is in contradiction with the argument of Imry and Ma described in Section 2.1. Moreover, Imbrie (1984) and Bricmont and Kupiainen (1987) proved rigorously the existence of an ordered state at finite temperature for $D = 3$ (whereas the pure Ising has *no* ordered state in $D = 1$), and Aizenman and Wehr (1989) finally proved that in $D \leq 2$ only one pure Gibbs state exists.

The flaw in the supersymmetry argument can be traced to the fact that Eq. (2.28) turns out to have many solutions, while it was already noticed by Parisi and Sourlas that their elegant proof is valid only when the solution is unique. In principle, to get the correct result, different solutions should be considered with different weights (Parisi, 1984). Dimensional reduction continues to hold for systems like 'branched polymers' (Parisi and Sourlas, 1981; Bridges and Imbrie, 2003).

To understand where perturbation expansion (and the renormalization group) have failed, we take a new look at it (Brezin and De Dominicis, 1998, 1999, 2001).

2.5 Generalized couplings

In deriving Eq. (2.7) we have implicitly replaced $h_i + \phi_i$ by ϕ_i and then expanded $\ln\cosh(\phi_i)$ in powers of ϕ_i, as is customary for small ϕ_i (we come from the paramagnetic side). However, h_i varies between $\pm\infty$, and $\phi_i + h_i$ cannot always be taken as small. So we start again from

$$\overline{Z^n} \simeq \int \prod_i (\mathrm{d}h_i) \frac{\mathrm{e}^{-\sum_i h_i^2/2\Delta}}{(2\pi\,\Delta)^{1/2}} \int \prod_{i,a} (\mathrm{D}\phi_i^a)$$
$$\times \exp\left\{ -\frac{1}{2\beta} \sum_{a,i,j} \phi_i^a (J^{-1})_{ij}\, \phi_j^a + \sum_{a,i} \ln\cosh\left(\phi_i^a + h_i\right) \right\} \tag{2.33}$$

where we have rescaled h_i. Expanding in ϕ_i^a, we have

$$\ln \frac{\cosh(\phi + h)}{\cosh h} = \ln\left(1 + \tau\phi + \frac{1}{2}\phi^2 + \frac{\tau}{6}\phi^3 + \frac{\phi^4}{24} + \cdots\right), \tag{2.34}$$

where τ stands for $\tanh(h)$. When we do the h-averaging we obtain quadratic and quartic terms in ϕ, as

$$\exp\left\{\sum_a \ln\left(\frac{\cosh(\phi^a+h)}{\cosh h}\right)\right\}$$
$$=\exp\left\{\frac{1}{2}(1-\tau_2)\,\sigma_2+\frac{\tau_2}{2}\sigma_1^2-\frac{1}{12}(1-4\tau_2+3\tau_4)\,\sigma_4-\frac{1}{3}(\tau_2-\tau_4)\,\sigma_1\sigma_3\right.$$
$$\left.+\frac{1}{8}(\tau_4-\tau_2^2)\sigma_2^2-\frac{1}{4}(\tau_4-\tau_2^2)\sigma_1^2\sigma_2-\frac{1}{24}(3\tau_2^2-\tau_4)\sigma_1^4\right\}, \tag{2.35}$$

where

$$\tau_p=\frac{1}{\sqrt{2\pi\Delta}}\int dh e^{-h^2/2\Delta}(\cosh h)^n(\tanh h)^p \tag{2.36}$$

and

$$\sigma_k=\sum_a(\phi^a)^k. \tag{2.37}$$

(Note that in this way $\tau_2=\Delta-2\Delta^2+O(\Delta^3)$ and $\tau_4=3\Delta^2-20\Delta^3+O(\Delta^4)$.)

Putting all together we now obtain, instead of (2.7), a new effective Lagrangian, where instead of the sole four-field coupling $V_1(\phi^a)=\frac{u_1}{4!}\sigma_4\equiv\frac{u_1}{4!}\sum_{a,i}(\phi_i^a)^4$ we have now the vertices

$$V(\phi^a)=\frac{u_1}{24}\sigma_4+\frac{u_2}{6}\sigma_1\sigma_3+\frac{u_3}{8}\sigma_2^2+\frac{u_4}{4}\sigma_1^2\sigma_2+\frac{u_5}{24}\sigma_1^4. \tag{2.38}$$

So our problem is now to evaluate

$$\overline{Z^n}=\int \mathrm{D}\phi_i^a \exp\left\{\mathcal{L}\{\phi_i^a\}+\sum_{a,i}H_i^a\phi_i^a\right\} \tag{2.39}$$

with

$$\mathcal{L}\{\phi_i^a\}=-\frac{1}{2}\sum_{ab}\int\frac{\mathrm{d}^Dp}{(2\pi)^D}\phi^a(p)\left[G^0(p)\right]_{ab}^{-1}\phi^b(-p)-\sum_i V(\phi_i^a). \tag{2.40}$$

Note that the five quartic terms appearing in (2.38) are all the quartic interactions compatible with the symmetry of the Lagrangian under replica permutations. With the exception of σ_4 restricted to a single replica, all these terms involve several replicas.

Let us now proceed with the perturbative treatment. It is important to carry on the symmetry weights in V, so that the weights of perturbation expansion can then be trivially read off graphs, provided one uses a faithful vertex representation, for example:

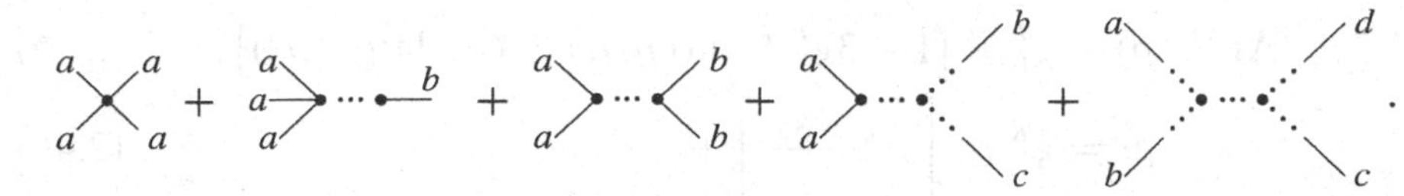

(2.41)

The dotted lines are to be thought of as dissolving if one removes the h-averaging (just like the disconnected component of the propagator separating out into two pieces).

Notice that all these couplings are *repulsive* (positive), except for u_3 which is *attractive*, and obviously can be viewed as resulting from the average over a random, quenched, temperature (r_0) term.

Our purpose now is to reexamine what the renormalization group also says on the Curie line. We shall keep to one loop, and thus need only examine what happens to the couplings (field strength renormalization occurs only at two loops with a ϕ^4 coupling).

2.6 Renormalization at one loop: single replica

We proceed now with the standard renormalization procedure (see Appendix A), namely: (i) define renormalized coupling constants to incorporate the UV divergences appearing in $\Gamma^{(4)}$ below the critical dimension; (ii) compute the beta functions; (iii) analyze the RG flow, find the fixed points and test for their stability.

Let us start for the moment with the simplified Lagrangian (2.10), where only one coupling constant, u_1, appears.

From Expression (2.22) we get, at first order,

$$\Gamma_1^{(4)}(p) = u_1 - 3u_1^2\Delta \int_0^{\Lambda} \frac{d\overset{D}{q}}{(2\pi)^D} \frac{1}{q^4(p+q)^2} + \cdots , \tag{2.42}$$

where $p = p_1 + p_2$. Λ is a UV cut off and the momentum integral still exhibits UV sensitivity, to be cured by an appropriate renormalization of the coupling constant. In fact, it is clear that the appropriate coupling is Δu_1 (and not u_1). Indeed, with one Δ per loop, we have $\Gamma_1^{(4)} \sim u_1(\Delta u_1)^{V-1}$, i.e. $\Delta\Gamma_1^{(4)} \backsim (\Delta u_1)^V$. It is important to signal that Δ will *not* renormalize. Setting $\Delta u_i \equiv g_i$, we have,

$$\Delta\Gamma_1^{(4)}(p) = g_1 - 3g_1^2 \int_0^{\Lambda} \frac{d\overset{D}{q}}{(2\pi)^D} \frac{1}{q^4(p+q)^2} + \cdots \tag{2.43}$$

and, at the symmetry point (see Appendix A),

$$\Delta\Gamma_1^{(4)}(p)\,|_{\mathrm{SP}} \equiv g_1^R\, \mu^{\varepsilon}, \tag{2.44}$$

where now $\varepsilon = 6 - D$. Expressing (2.43) in terms of g_1^R, we get

$$\Delta\Gamma_1^{(4)}(p) = g_1^R \mu^\varepsilon \left[1 - 3g_1^R S_D \ln(p/\mu) + O(\varepsilon \ln^2(p/\mu))\right], \tag{2.45}$$

$$g_1 = g_1^R \mu^\varepsilon \left[1 + \frac{3g_1^R}{\varepsilon}\right] + \cdots \tag{2.46}$$

Hence, the beta function is

$$\beta_1(g_1^R) = -\varepsilon g_1^R + 3\left(g_1^R\right)^2, \tag{2.47}$$

which has a fixed point at $g_1^{R*} = \frac{\varepsilon}{3}$, and the *flow* given by the same equation as for the pure system (see Appendix A) with the exponent $\omega = \beta'(g_1^{R^*}) = +\varepsilon$ unchanged (ε understood as $4 - D$, $6 - D$ respectively). Hence both the fixed point and the exponent ω satisfy *dimensional reduction*, as expected after the discussion in the preceding sections. Note that the single replica calculation leads to a stable result. As we shall see, when taking care of the other couplings, this will no longer be true.

2.7 Renormalization at one loop: multi-replicas

We want now to take into account the generalized couplings appearing in (2.38). A convenient algebraic way to evaluate the one-loop contribution to the N-point vertex function is to look at the one-loop expression of the Gibbs free energy $\Gamma\{M_i^a\}$:

$$-\Gamma\left\{M_i^a\right\} = \left\{\mathcal{L}\left\{\phi_i^a\right\} + \frac{1}{2}\mathrm{tr}\ \ln\left(-\partial^2\mathcal{L}\left\{\phi_i^a\right\}/\partial\phi_i^a\,\partial\phi_j^b\right)\right\}\Bigg|_{\phi_i^a = M_i^a}. \tag{2.48}$$

Indeed, $\Gamma\{M_i^a\}$ is the generating functional for the N-point vertex functions and, developing the tr log in (2.48), we read directly the Nth contribution. However, with five distinct couplings the computation may require some attention. To avoid being drowned in the algebra, we keep to a simplified model with the three key-couplings u_1, u_2 and u_3. This will be enough to make our point. In the end we shall mention how the results generalize. Then, instead of (2.38), we restrict V to

$$V(\phi^a) = \frac{u_1}{24}\sum_a(\phi^a)^4 \quad + \quad \frac{u_2}{6}\sum_a(\phi^a)^3\sum_b\phi^b \quad + \quad \frac{u_3}{8}\left[\sum_a(\phi^a)^2\right]^2 \tag{2.49}$$

From (2.48) we get, to one loop:

$$\frac{1}{8}\Gamma_3^{(4)}$$

$=$ 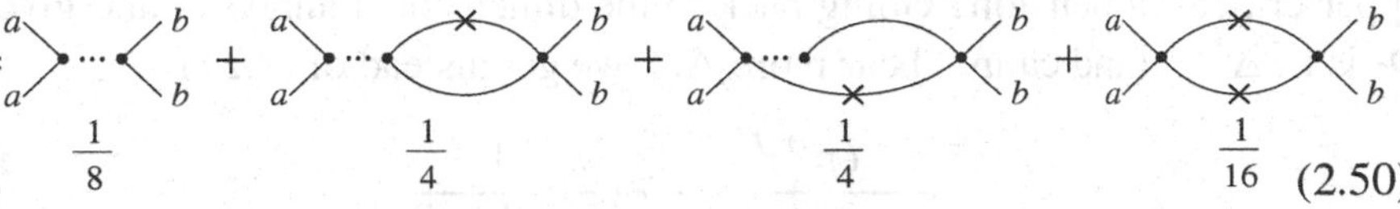 (2.50)

and

$\frac{1}{6}\Gamma_2^{(4)} =$ 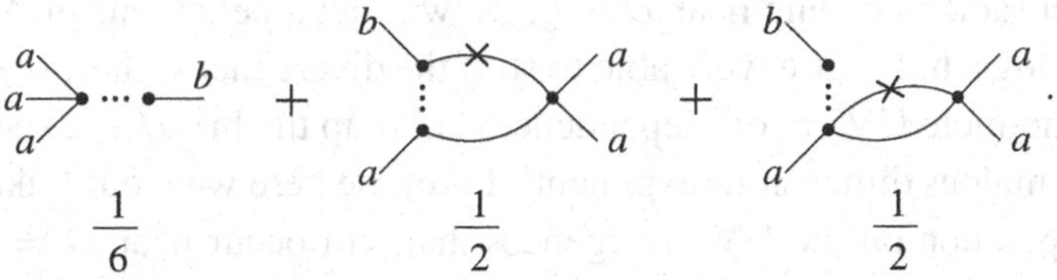 (2.51)

In algebraic terms, setting $g_i = \Delta u_i$, we get

$$\Delta\Gamma_3^{(4)}(p) = g_3 - 2\,(g_3 + g_2)\,\frac{g_1}{\varepsilon p^{\varepsilon}} - \frac{1}{2}\,\frac{\Delta}{p^{2+\varepsilon}} g_1^2, \tag{2.52}$$

$$\Delta\Gamma_2^{(4)}(p) = g_2 - 3\,(g_3 + g_2)\,\frac{g_1}{\varepsilon p^{\varepsilon}}\,. \tag{2.53}$$

The new fact is now the emergence of a strong infrared (IR) *singular contribution* to $\Gamma_3^{(4)}(p)$ (for $\Gamma_2^{(4)}$ an analogue term only occurs at two loops). Let us now look at Eqs. (2.52) and (2.53) and proceed in steps to get the RG flow.

2.7.1 General structure of the $\Gamma^{(4)}$ functions

It can be shown (Brezin and De Dominicis, 2001) that the structure of $\Gamma_3^{(4)}$ is

$$\Delta\Gamma_3^{(4)} = g_3 + (g_3 + g_2)\,C_3\,(g_1) + D_3\,(g_1)\,. \tag{2.54}$$

Here $C_3\,(g_1)$ is *connected*, with one Δ per loop only. $D_3(g_1)$ is *disconnected* (upon removal of the h-averaging), it bears one Δ per loop plus an extra one and is strongly infrared divergent as in (2.52).

Likewise one gets for $\Gamma_2^{(4)}$

$$\Delta\Gamma_2^{(4)} = g_2 + (g_2 + g_3)\,C_2(g_1) + D_2(g_1) \tag{2.55}$$

except that D_2 starts at two loops, i.e. with a term in $(\Delta/p^{2+2\varepsilon})\,g_1^3$.

The structures described by (2.54) and (2.55) are easily verified by recurrence.

2.7.2 IR divergences

What should we do with the strong IR divergent terms of D_3? Of course with a $D_3 \sim 1/p^{(8-D)}$ the first question is: should we not consider rather $D = 8$ as the upper critical dimension? Going back to the dimensional analysis, and given that D_3 is in Δ^{L+1} (and cannot bear more Δs), we get instead of (2.27)

$$\frac{(p^D)^L}{(p^2)^{2L}(p^2)^{L+1}} \sim \frac{1}{p^2 p^{L(6-D)}}. \tag{2.56}$$

Thus, near $D = 8$ we won't get a build-up of logarithms (like in (2.27) or in C_3 near $D = 6$), but rather, keeping near $D = 6$, we will get a behaviour in Δ/p^2 times a build-up of logs. In $\Gamma_1^{(4)}$ we were able to stuff the divergences (the $1/\varepsilon$) into u_1^R and then (with no more UV cut off dependence) build up the $\ln(p/\mu)$ dependence into the $\Gamma^{(4)}$ anomalous dimension (exponent). Likewise here we would like to perform the same operation on the UV divergences that will occur near $D = 6$, and if so, which renormalized quantity is now going to do the job?

The answer is that what has been done for the renormalization of g_1 is enough to take care of $D_3(g_1)$. Indeed, that can be seen in a pedestrian way by looking at two-loops, and one gets

$D_3\,(p; g_1) =$ 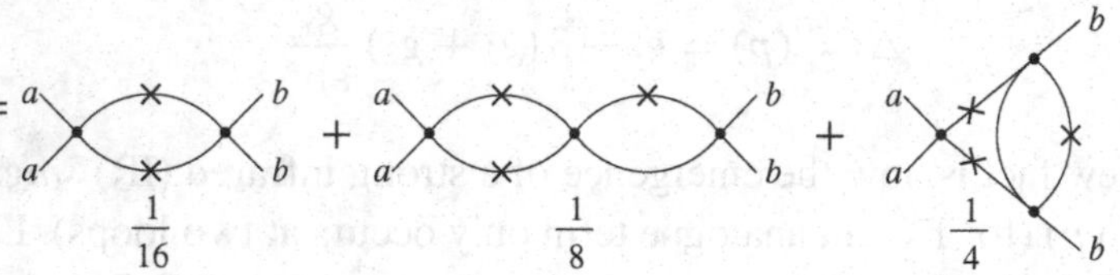

$$= -\frac{1}{2}\frac{g_1^2\Delta}{p^{2+\varepsilon}}\left\{1 - \frac{2g_1}{p^{\varepsilon}} - \frac{4g_1}{p^{\varepsilon}} - \cdots\right\}, \tag{2.57}$$

which, using (2.46), is equal to

$$-\frac{1}{2}\frac{\Delta}{p^2}\left(g_1^R\right)^2 \mu^{\varepsilon}\left(\frac{\mu}{p}\right)^{\varepsilon}\left\{1 + \frac{6g_1^R}{\varepsilon} - \frac{6g_1^R}{\varepsilon}\left(\frac{\mu}{p}\right)^{\varepsilon}\right\}. \tag{2.58}$$

Thus

$$D_3\left(p; g_1^R\right) = -\frac{1}{2}\frac{\Delta}{p^2}\left(g_1^R\right)^2 \mu^{\varepsilon}\left(\frac{\mu}{p}\right)^{\varepsilon}\left\{1 + 6g_1^R \ln\left(p/\mu\right) + \cdots\right\}, \tag{2.59}$$

which is now renormalized, through the use of the relationship obtained at the $\Gamma_1^{(4)}$ level between g_1 and g_1^R. The logs build up an anomalous exponent as they did in $\Gamma_1^{(4)}$. The IR singular function D_3 obeys a Callan–Symanzik equation just as the standard four-point function did.

2.7.3 RG flow equations and stability on the Curie line

Having the more IR singular part of $\Gamma_1^{(4)}$ under control, we may now treat the C_3 part. There we have a very simple behaviour, since it will suffice to decide that g_2^R and g_3^R take care of whatever divergence (i.e. powers of $1/\varepsilon$) is left after use of the replacement of g_1 by its series in g_1^R. Expressing the μ-independence of the bare g_3 (or g_2), i.e. via $\mu\, \mathrm{d}/\mathrm{d}\mu \equiv \mu\, \partial/\partial\mu + \beta_{2,3}\, \partial/\partial g_{2,3} = 0$ (see Appendix A), together with (2.47), we get the complete set of beta functions:

$$\beta_1 = -\varepsilon g_1^R + 3\left(g_1^R\right)^2, \tag{2.60}$$

$$\beta_2 = -\varepsilon g_2^R + 3\left(g_2^R + g_3^R\right) g_1^R, \tag{2.61}$$

$$\beta_3 = -\varepsilon g_3^R + 2\left(g_2^R + g_3^R\right) g_1^R. \tag{2.62}$$

The fixed points of the RG flow are given by the set of equations $\beta_i = 0$. It is easy to check that the only nontrivial fixed point is the one that is obtained when only g_1^* is nonvanishing, i.e. $g_1^* = \varepsilon/3$, $g_2^* = g_3^* = 0$, which is the result one gets with only the simple u_1 vertex (see previous section). However, things appear differently when looking at the stability of the fixed point. Now the space of the couplings is three dimensional, and the stability of a fixed point is given by the eigenvalues of the matrix

$$\frac{\partial \beta_i}{\partial g_j} \equiv \Omega_{ij} \tag{2.63}$$

evaluated at that point. In the present case the eigenvalues λ are given by

$$0 = \det \Omega_{ij}(\lambda) = \begin{vmatrix} -\lambda - \varepsilon + 6g_1^* & 0 & 0 \\ 0 & -\lambda - \varepsilon + 3g_1^* & 3g_1^* \\ 0 & 2g_1^* & -\lambda - \varepsilon + 2g_1^* \end{vmatrix} \tag{2.64}$$

and are

$$\lambda_1 = +1, \quad \lambda_2 = \frac{2}{3}, \quad \lambda_3 = -1. \tag{2.65}$$

Hence the system is *unstable* on the Curie line. Note that the instability $\lambda_3 = -1$ originates from the g_3 coupling. So, contrary to the single replica approach, we are confronted here with an explicit instability.

2.8 A vitreous transition

From the above discussion it follows that we are left with an effective coupling between a pair of distinct replicas a, b that is both *attractive* and *singular*:

$$\Delta\Gamma^{(4)}(p) \sim -\frac{\Delta}{p^{2+\varepsilon}} g_1^2, \tag{2.66}$$

so that it is most natural to ask whether that could give rise to a bound state for a pair of fields $\phi^a\phi^b$ conjugate to the 'external' field Δ. Numerical indications that this is indeed the case can be found in Parisi and Sourlas (2002). The physical interpretation of this phenomenon is related to the presence of different pure states and implies the occurrence of a vitreous transition. The crucial issue is therefore to detect and discuss this transition for the RFIM. Mézard and Young (1992) and Mézard and Monasson (1994), using a self-consistent $1/m$ expansion on an m-component version of the RFIM, have argued that the singular behaviour of the two-point functions on the Curie line is due to the pre-occurrence of a vitreous transition, before reaching the Curie line. However, other arguments have been proposed suggesting that the correlation length associated with the ferromagnetic order is always larger than or equal to the correlation length that becomes critical at the vitreous transition (Parisi and Sourlas, 2003; Parisi, private communication, 2005), thus putting the transition to the glassy phase below or at the Curie line.

Here we wish just to make a few general remarks.

2.8.1 Legendre Transforms

Let us open a parenthesis and consider the general Lagrangian

$$\mathcal{L}\{\phi_i\} = \sum_i H_i\phi_i + \frac{1}{2}\sum_{ij} V_{ij}\phi_i\phi_j - \frac{u_1}{4!}\sum_i \phi_i^4. \tag{2.67}$$

As a first step, let us take $V_{ij} = -[G^0]^{-1}_{ij}$, then (after Fourier transforming the quadratic term) (2.67) becomes as (2.7), with the external source H_i replacing the random field h_i.

Let us consider the Legendre Transform constructed by

$$\Gamma\{M_i\} = F\{H_i\} + \sum_i H_i M_i, \tag{2.68}$$

where by definition, with $\exp(-F) = Z$,

$$-\frac{\partial F}{\partial H_i} = \langle\phi_i\rangle = M_i, \tag{2.69}$$

and thus, from (2.68),

$$\frac{\partial \Gamma}{\partial M_i} = H_i. \tag{2.70}$$

The functional $\Gamma\{M_i\}$ is easily derived as given by all connected graphs with no external lines, built with G^0 lines and vertices u_1, but such that the graphs do not fall into two pieces by cutting *one* line. Equation (2.70) is then the equation of motion determining the magnetization (induced, if $H_i \neq 0$, spontaneous otherwise).

The Jacobian of the transformation (from F to Γ),

$$\frac{\partial H_i}{\partial M_j} \equiv \frac{\partial^2 \Gamma}{\partial M_i \, \partial M_j}, \tag{2.71}$$

signals the transition through the occurrence of a zero eigenvalue for $M_i = 0$. In the low temperature phase, where ergodicity is broken, the two existing pure states are characterized by different values of the spontaneous magnetization. The zero eigenvalue of the Jacobian at the transition indicates that at T_c the value of the magnetization is no longer well defined: below T_c, to compute the equilibrium magnetization one needs to select by hand one of the two pure states, for example by applying an infinitesimal magnetic field. Alternatively, the inverse matrix of (2.71), the susceptibility

$$\chi_{ij} = -\frac{\partial^2 F}{\partial H_i \, \partial H_j}, \tag{2.72}$$

acquires an infinite eigenvalue (at zero momentum) at T_c.

Let us consider now a second Legendre Transform (for a paramagnetic system with $H_i = 0$ and $M_i = 0$). We construct

$$\Gamma\{G_{ij}\} = F\{V_{ij}\} + \frac{1}{2} \sum_{ij} V_{ij} G_{ij} \tag{2.73}$$

with, again, by definition,

$$-\frac{\partial F}{\partial V_{ij}} = \langle \phi_i \phi_j \rangle = G_{ij}, \tag{2.74}$$

and thus, from (2.73),

$$\frac{\partial \Gamma}{\partial G_{ij}} = V_{ij}. \tag{2.75}$$

The functional $\Gamma\{G_{ij}\}$ was first given by Luttinger and Ward (1960),

$$\Gamma\{G_{ij}\} = -\frac{1}{2} \mathrm{tr} \ln G + \mathcal{K}^{(1)}\{G\}, \tag{2.76}$$

where $\mathcal{K}^{(1)}$ is given by all the graphs, connected with no external lines, built with G lines and $(u_1/4!)\sum_i \phi_i^4$ vertices, such that they do not fall into two pieces if one cuts *two* lines. Equation (2.75) is then the equation of motion determining G. For $V_{ij} = -[G^0]_{ij}^{-1}$ (i.e. no external conjugate source) we recover Dyson's equation.

The Jacobian of the transformation becomes the matrix

$$\frac{\partial V_{ij}}{\partial G_{kl}} = \frac{\partial^2 \Gamma}{\partial G_{ij} \partial G_{kl}}. \tag{2.77}$$

When an eigenvalue of (2.77) reaches zero, it signals the occurrence of a bound state. More about that later.

So let us return now to the *RFIM* system, where the field ϕ_i is replaced by ϕ_i^a, the magnetization M_i by M_i^a and the Lagrangian $\mathcal{L}\{\phi_i\}$ of (2.67) by $\mathcal{L}\{\phi_i^a\}$ of (2.16). We write now the quadratic term in $\mathcal{L}\{\phi_i^a\}$ as V_{ij}^{ab}, and in the end, setting the external source to zero, we have

$$V^{ab}(p) \to -\left[G^0\right]_{ab}^{-1}(p) \tag{2.78}$$

with $[G^0]_{ab}^{-1}(p)$ given by (2.11). The generalized functional for the *double* Legendre Transform can be constructed (De Dominicis and Martin, 1964), but to keep things simple we stay above (or at) the Curie line where $M_i^a = 0$. Then one can make with $\Gamma\{G_{ij}^{ab}\}$ as in (2.76) where $\mathcal{K}^{(1)}$ is one-irreducible with respect to G_{ab} lines. More precisely, we write

$$G_{ab} \equiv G_a \delta_{a;b} + C_{ab}, \tag{2.79}$$

where we have separated the connected contribution G_a, and the disconnected one, C_{ab}, that becomes h-connected under h-averaging (like in (2.13)). Re-expressing Γ in terms of components, we obtain

$$-\Gamma\{G_a, C_{ab}\} = \frac{1}{2}\mathrm{tr}\ln\left(G_a\delta_{a;b} + C_{ab}\right) + \sum_{s=1} \mathcal{K}_s^{(1)}\{G_a, C_{ab}\}, \tag{2.80}$$

where we have separated the one-irreducible contributions according to their number of free replica indices (after account of the $\delta_{a;b}$ constraint). For example,

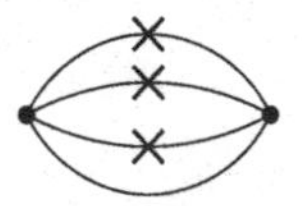

belongs to $s = 1$,

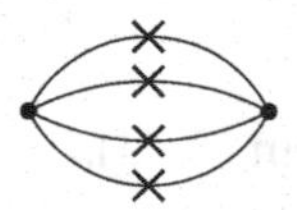

belongs to $s = 2$.

At this point, given that we are going to search for zero eigenvalues of some Jacobian, we would like to understand which order parameter should characterize the vitrous phase. Consider the system below the vitrous transition and above the Curie line: many pure states exist, all with global magnetization $M = 0$, what distinguishes them? As we have already discussed, the random field h_i induces a nonzero local magnetization $\langle\phi_i\rangle$. If many states exist, the local magnetization will be different in each one of them. Given two copies of the system a, b, we expect each copy to be in a pure state, but not necessarily the same, and in general $\langle\phi_i^a\rangle \neq \langle\phi_i^b\rangle$. However, when averaging over the random field, the averaged local magnetization $\overline{\langle\phi_i^a\rangle} = 0$ vanishes for every a, and thus is not an appropriate order parameter. On the other hand, the average local magnetization correlator $C_{ab} = \overline{\langle\phi_i^a\rangle\langle\phi_i^b\rangle}$, which is proportional to Δ at the bare level, may assume different values with varying a and b, reflecting the spectrum of correlations existing between the pure states. Therefore, if many pure states exist and ergodicity is broken, we expect C_{ab} to depend in a nontrivial way on the two replica indices. The order parameter of our problem is then the field-connected part C_{ab} of the two-point correlation function.

2.8.2 The eigenvalues of the Jacobian

The Jacobian matrix

$$\frac{\partial^2\Gamma}{\partial C_{ab}(p)\,\partial C_{cd}(p')} \equiv \mathcal{M}_{ab;cd}(p;p') \tag{2.81}$$

can be calculated from (2.80) and its zero eigenvalues will determine the (vitreous) transition. The tensor $\mathcal{M}$ bears four replica indices and its diagonalization in the replica space is not a priori a simple task. However, since we are approaching the transition line from above, we come from the paramagnetic phase where only a pure state exists (the paramagnetic one), and the structure of the propagator is simpler, namely $C_{ab} = C_{aa} = C$. The dependence in replicas can then be block-diagonalized quite easily by distinguishing different sectors in the replica space, with the techniques that are discussed in detail in the next chapters. In the *so-called replicon sector*, which is the most sensitive to the transition, the structure of $\mathcal{M}$ is the one of a simple matrix in (p, p'), e.g., keeping the lowest order (attractive) term in g_1^2, we have

$$\mathcal{M}_R(p;p') = G^{-1}(p)\,G^{-1}(p') - \mathcal{K}_2^{(1)}(p;p'), \tag{2.82}$$

where only $s = 2$ terms contribute, with

$$\mathcal{K}_2^{(1)}(p;p') \quad = \quad [\text{diagram with external legs } p,\ p',\ -p,\ -p'] \; . \tag{2.83}$$

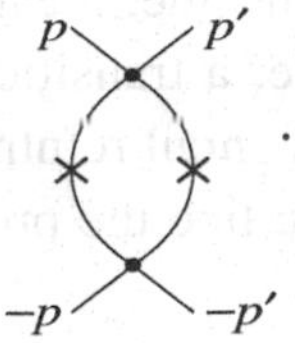

Had we chosen to investigate $F\{\Delta\}$ we would have ended up with

$$-\frac{\partial^2 F}{\partial \Delta_{ab}(p)\,\partial \Delta_{cd}(p')} = (\mathcal{M}^{-1})_{ab;cd} \equiv \mathcal{G}^{ab;cd}(p;p'), \qquad (2.84)$$

which, as above, results now into

$$\mathcal{G}_R(p;p') = G^2(p)\left[\delta_{p+p';0} + \sum_{p''} \mathcal{K}_2^{(1)}(p;p'')\mathcal{G}_R(p'';p')\right], \qquad (2.85)$$

that is, in terms of diagrams,

$$\mathcal{G}_R(p;p') = \quad + \quad + \quad + \cdots$$

$$= \left[G^{-2} - \mathcal{K}_2^{(1)}\right]^{-1}_{p;p'} = \left[\mathcal{M}_R\right]^{-1}_{p;p'}. \qquad (2.86)$$

A zero eigenvalue of $\mathcal{M}_R(p;p')$ would be seen then as a pole (or a cut) in $\mathcal{G}_R(p;p')$. Note that $\mathcal{G}_R$ is directly related to a simple susceptibility. If we define

$$\mathcal{X}_R(r;r') \equiv \overline{\left[\langle \phi(r)\,\phi(r')\rangle - \langle \phi(r)\rangle\,\langle \phi(r')\rangle\right]^2}, \qquad (2.87)$$

we then have

$$\sum_{p;p'} \mathcal{G}_R(p;p') = \sum_{r;r'} \mathcal{X}_R(r;r'). \qquad (2.88)$$

2.8.3 *A naive estimate*

Equations (2.82) and (2.83) hold above the vitreous transition, and should exhibit a zero eigenvalue at the transition. Unfortunately, we are not able to solve the associated eigenvalue equation to determine exactly the location of the transition. However, we can at least give a crude estimate of it.

If we sit on the Curie line and the supposed vitreous transition lies above it (as suggested by Mézard and coworkers), then (2.82) is not the correct form for $\mathcal{M}$, since to get it we have assumed replica independence in the propagator, which breaks down in the vitreous phase. However, this fact should be signalled by the presence of a *negative* eigenvalue, meaning that the solution we are looking at is thermodynamically unstable, a transition has already occurred and a removal of negative eigenvalues would entail reintroduction of some replica dependence. On the other hand, on the Curie line the propagators are massless and the eigenvalue computation simplifies.

Writing the eigenvalue equation associated with (2.82), we have

$$G^{-2}(p)\, f_\lambda(p) - g_1^2 \int \frac{\mathrm{d}^D q}{(2\pi)^D}\, C_2(q)\, f_\lambda(p-q) = \lambda f_\lambda(p), \tag{2.89}$$

with

$$C_2(q) = \int C^2(r)\, \mathrm{e}^{\mathrm{i}q\cdot r} \mathrm{d}^D r. \tag{2.90}$$

On the Curie line we may take $G(p) \sim 1/p^{2-\eta}$ and $C(p) \sim 1/p^{4-\bar{\eta}}$, η and $\bar{\eta}$ being the two (a priori different) anomalous dimensions. To find the lowest eigenvalue of the above 'Schrödinger' equation (with an almost ∇^4 kinetic term) we resort to the Rayleigh–Ritz variational method which provides an *upper bound*, by using a normalized trial wave function $f_\lambda(p)$ whose parameters are determined variationally, and we find

$$\lambda = \int \frac{\mathrm{d}^D q}{(2\pi)^D}\, p^{4-2\eta} f^2(p) - g_1^2 \int \frac{\mathrm{d}^D q}{(2\pi)^D}\, C_2(q)\, \phi(q),$$

$$\phi(q) = \int \frac{\mathrm{d}^D q}{(2\pi)^D}\, f(p)\, f(p-q), \qquad \phi(0) = 1.$$

Scaling out the unit length R, we get

$$\lambda = \frac{a}{R^{4-2\eta}} - \frac{g_1^2 b}{R^{2D-8+2\bar{\eta}}}. \tag{2.91}$$

Stationarity with respect to R^2 yields

$$0 = \frac{(2-\eta)\, a}{(R^2)^{2-\eta}} - \frac{(D-4+\bar{\eta})\, g_1^2 b}{(R^2)^{D-4+\bar{\eta}}}, \tag{2.92}$$

hence

$$\lambda = (D-6+\bar{\eta}+\eta)\, \frac{g_1^2}{(2-\eta)\,(R^2)^{D-4+\eta}}. \tag{2.93}$$

So the *eigenvalue upper bound remains negative (on the Curie line) below $D = 6$*. This seems to indicate that, before we reach the Curie line, the vitreous transition has occurred, with an order parameter $C_{ab}(p)$. To remove the occurrence of negative eigenvalues, one should then resort to allowing a dependence upon a, b, or rather to what will be termed later the overlap of a, b. In other words one should then resort to *Replica Symmetry Breaking* (RSB).

In the above we have assumed that the vitreous transition occurs above the Curie line and checked it at the one-loop level (for the Jacobian). The effect of higher loop contributions may change the answer (Parisi, private communication, 2005).

In that case, one would have to start the calculation again, now with the double functional $\Gamma\{M_i^a, G_{ab}(p)\}$ to decide whether the vitreous phase sets in at or below the Curie line.

2.9 Summary

In this chapter we have studied the statics of the RFIM using a perturbative approach. We have shown that:

- The effective Lagrangian with only the single replica $\sum_a \sum_i (\phi_i^a)^4$ interaction gives dimensional reduction, and therefore fails to correctly describe the physics of the RFIM, for which the existence of a transition in $D = 3$ has been proved by other means.
- A proper analysis starting from the original spin Hamiltonian shows that one should consider also multi-replica interactions in the effective Lagrangian. Considering these new couplings, one discovers that the fixed point corresponding to dimensional reduction is actually unstable.
- The RG flow obtained at one loop with the complete Lagrangian (including the multi-replica couplings) does not have a stable fixed point. Thus, either a stable fixed point which is *not* of order ε exists, or the transition is first order. In any case the epsilon-expansion of the critical properties of the RFIM is not possible, and dimensional reduction breaks down at first order in ε.
- The analysis of the generalized Gibbs free energy $\Gamma(G_{ab})$ at one loop indicates that a vitreous transition to a replica symmetry broken phase may occur above the Curie line.

References

Aharony A., Imry T. and Ma S. K. A. (1976). *Phys. Rev. Lett.*, **37**, 1364.
Aizenman M. and Wehr J. (1989). *Phys. Rev. Lett.*, **62**, 2503.
Belanger D. P. (1998). In *Spin Glasses and Random Fields*, ed. A. P. Young, Singapore, World Scientific, p. 251.
Brezin E. and De Dominicis C. (1998). *Europhys. Lett.*, **44**, 13.
Brezin E. and De Dominicis C. (1999). *C. R. Acad. Sci. II*, **327**, 383.
Brezin E. and De Dominicis C. (2001). *Eur. Phys. J. B*, **19**, 467.
Bricmont J. and Kupiainen A. (1987). *Phys. Rev. Lett.*, **59**, 1829.
Bridges D. C. and Imbrie J. Z. (2003). *Ann. Math.*, **158**, 1019; *J. Statist. Phys.*, **110**, 503.
Cardy J. (1984). *Phys. Rev. B*, **29**, 505.
De Dominicis C. and Martin P. (1964). *J. Math. Phys.*, **118**, 31.
Fishman S. and Aharony A. (1979). *J. Phys. C*, **12**, L729.
Imbrie J. Z. (1984). *Phys. Rev. Lett.*, **53**, 1747.
Imry J. Z. and Ma S. K. (1975). *Phys. Rev. Lett.*, **35**, 1399.
Luttinger J. and Ward J. (1960). *Phys. Rev.*, **118**, 1417.
Mézard M. and Monasson R. (1994). *Phys. Rev. B*, **50**, 7199.
Mézard M. and Young A. P. (1992). *Europhys. Lett.*, **18**, 653.
Natterman T. (1998). In *Spin Glasses and Random Fields*, ed. A. P. Young, Singapore, World Scientific, p. 277.

Parisi G. (1984). In les Houches Session XXXIX, 1982, *Recent Advances in Field Theory and Statistical Mechanics*, eds. J.-B. Zuber and R. Stora, Amsterdam, Elsevier Science Publishers B.V.
Parisi G. and Sourlas N. (1979). *Phys. Rev. Lett.*, **43**, 744.
Parisi G. and Sourlas N. (1981). *Phys. Rev. Lett.*, **46**, 871.
Parisi G. and Sourlas N. (2002). *Phys. Rev. Lett.*, **89**, 257204.
Tarjus G. and Tissier M. (2003). *Phys. Rev. Lett.*, **93**, 267008.
Young A. P. (1976). *Phys. Rev. Lett.*, **37**, 944.
Zinn-Justin J. (1989). *Quantum Field Theory and Critical Phenomena*, Oxford, Clarendon Press, Chapter 29.

3

The dynamical approach

The replica trick leaves an unsatisfactory taste, on account of its unphysical character. We will now introduce an alternative, more physical approach, based on dynamics (De Dominicis, 1978). Recently, the dynamical approach has proved extremely powerful in addressing the complex behaviour of glassy systems at low temperature (see Bouchaud *et al.*, 1998, for a review of recent results). For mean field spin glass systems the dynamics has been fully solved, revealing interesting new patterns of off-equilibrium behaviour where time translational invariance is broken and fluctuation–dissipation relations are violated. Interestingly, the theoretical framework developed for mean field systems seems to be an adequate starting point to describe a great variety of finite dimensional systems, from real spin glasses to glass forming liquids, driven granular matter, critical and coarsening systems, etc.

In this chapter, we describe the dynamical approach from a general perspective, building the formalism step by step, introducing the dynamical observables and their properties. Our starting point is the Langevin equation, that describes the relaxational dynamics of a system in contact with a thermal bath (Section 3.1). Then, for a Ginzburg–Landau–Wilson Hamiltonian, we illustrate how to perform a perturbative analysis via a tree-like expansion of the dynamical equation (Section 3.1.2), or through the more sophisticated Martin–Siggia–Rose generating functional (Section 3.2). Finally, the whole approach is generalized for the Random Field Ising Model where complications arise due to the presence of the quenched disorder.

For the time being we limit our analysis to the equilibrium dynamics, that is, we shall approach the vitreous phase from above. In this sense we do not treat what is probably the most interesting region where off-equilibrium dynamics with all its new intriguing behaviour occurs. The reason is that we are considering a finite dimensional model, with the complications associated with the loop expansion. In one of the next chapters we will use the alternative perspective, considering a simple mean field system and looking at its off-equilibrium behaviour.

3.1 Langevin dynamics

Given a field $\phi(x)$ with Hamiltonian $\mathcal{H}(\phi)$, e.g. as in (2.7), we consider its dynamical evolution as described by the following Langevin equation:

$$\frac{\partial\phi(x;t)}{\partial t} = -\frac{\partial\mathcal{H}(\phi)}{\partial\phi} + \eta(x;t). \tag{3.1}$$

Here $\eta(x;t)$ represents a fluctuating thermal noise controlled by its probability distribution (appropriately normalized)

$$\mathcal{P}\{\eta\} \sim \exp\left\{-\frac{1}{4D_0}\int \mathrm{d}x^D \int \mathrm{d}t\, \eta^2(x,t)\right\}, \tag{3.2}$$

i.e. we have

$$\begin{aligned} \langle\eta\rangle &= 0, \\ \langle\eta(x;t)\eta(x';t')\rangle &= 2D_0\delta^D(x-x')\delta(t-t'), \end{aligned} \tag{3.3}$$

the angle brackets meaning average over the probability distribution, and D_0 being the diffusion coefficient.

Originally the Langevin equation was introduced to describe Brownian motion, i.e. the time evolution of a small particle suspended in a viscous fluid at a given temperature (see, e.g., Gardiner, 1983; Parisi, 1988, Chapter 19). In that context, the field ϕ represents the coordinate of the Brownian particle, the deterministic gradient term in Eq. (3.1) describes the frictional force acting on it (with the friction coefficient set to one) and the noise describes the fluctuations due to the heat bath[†]. In this sense the Langevin equation was introduced as a phenomenological description of Brownian motion. In certain cases, however, it can be fully derived from the deterministic equation of motion (see, e.g., Zwanzig, 2001, Chapter 1): if we consider a system interacting in a simple way with a heat bath of harmonic oscillators, the equations of motion of the system become formally equivalent to a generalized Langevin equation where the friction and the noise term are some functions of the frequency distribution of the harmonic oscillators and the initial conditions of the bath. By choosing an adequate frequency spectrum and thermalized initial condition, the statistical properties of the noise term are appropriately described by a Gaussian random variable. Another important property of the Langevin equation concerns the field distribution $\mathcal{P}(\phi)$: it obeys the so-called Fokker–Planck equation and can be shown to converge asymptotically to the canonical distribution $\exp[-\mathcal{H}(\phi)/T]$, where T is the temperature, and $T = D_0$ (Einstein relationship).

[†] In the Langevin equation there is in principle also an inertial term; however, if one is interested only in time scales much longer than m/γ (m being the mass of the particle and γ the friction coefficient), this term can be discarded.

Thus, this equation describes the behaviour of a system (or of certain degrees of freedom), in contact with a thermal bath, that relaxes toward equilibrium.

Let us now proceed with our analysis considering the familiar ϕ^4 Hamiltonian,

$$\mathcal{H}(\phi) = \frac{1}{2} \int \frac{\mathrm{d}p^D}{(2\pi)^D}\, \phi(p) E_p \phi(-p) + \frac{u_1}{4!} \sum_{p_i} \phi(p_1)\phi(p_2)\phi(p_3)\phi(p_4)\delta^{Kr}_{\sum p;0} \tag{3.4}$$

with $E_p \equiv p^2 + r_0$ and where $\sum_{p_i}$ indicates integration over all the momenta p_i, $i = 1, \ldots, 4$. Equation (3.1) then becomes

$$\frac{\partial \phi(p;t)}{\partial t} = -\left[E_p \phi(p;t) + \frac{u_1}{6} \sum_{p_1 p_2} \phi(p_1;t)\phi(p_2;t)\phi(p - p_1 - p_2;t) \right] + \eta(p;t). \tag{3.5}$$

After integration we get

$$\begin{aligned} \phi(p;t) &= \phi_0(p)\, \mathrm{e}^{-E_p(t-t_0)} + \int_{t_0}^{t} \mathrm{d}s\, \mathrm{e}^{-E_p(t-s)} \eta(p;s) \\ &\quad - \frac{u_1}{6} \int_{t_0}^{t} \mathrm{d}s\, \mathrm{e}^{-E_p(t-s)} \sum_{p_1 p_2} \phi(p_1;s)\phi(p_2;s)\phi(p - p_1 - p_2;s), \end{aligned} \tag{3.6}$$

where we have used the Fourier transform of $\eta(x)$, i.e. $\eta(p)$, for convenience. That (3.6) is the solution of (3.5) with initial condition $\phi(p;t_0) \equiv \phi_0(p)$ is easily checked by differentiation. Note that (3.6) can be now solved by iteration, giving rise to a 'tree' (see Section 3.1.2).

3.1.1 The bare average values

Let us now look at the relevant quantities, averaged over the noise, in the simplest approximation, i.e. disregarding the ϕ^4 term.

- *Magnetization*

  Since $\langle \eta \rangle = 0$, we get from (3.6)

$$\langle \phi(p;t) \rangle = \phi_0(p)\, \mathrm{e}^{-E_p(t-t_0)}. \tag{3.7}$$

  Here we see that if $(t - t_0) \to +\infty$ (i.e. either $t_0 \to -\infty$ or $t \to +\infty$) the system loses memory of the initial conditions. We could also decide that the initial conditions are given by a distribution law, e.g. a Gaussian law $P(\phi_0) \sim \exp -\sum_p \phi_0^2(p)/2\phi_0^2$. This would lead to, taking an average over $\phi_0(p)$,

$$\langle\langle \phi(p;t) \rangle\rangle_{\phi_0} = 0 \tag{3.8}$$

just by parity conservation. In the following we shall mostly take $\phi_0(p) = 0$ for simplicity, but there is no difficulty of principle in keeping track of the initial conditions.

- *Correlation*
At the *bare* level, we get

$$\langle \phi(p;t)\phi(p';t')\rangle_0 = \delta^{kr}_{p+p';0}\ C_0(p;t,t') \tag{3.9}$$

with

$$\begin{aligned} C_0(p;t,t') &= \phi_0^2(p)\, \mathrm{e}^{-E_p(t+t'-2t_0)} \\ &\quad + 2T \int_{t_0}^{t} \mathrm{d}s\, \mathrm{e}^{-E_p(t-s)} \int_{t_0}^{t'} \mathrm{d}s'\, \mathrm{e}^{-E_p(t'-s')}\, \delta(s-s') \\ &= \phi_0^2(p)\, \mathrm{e}^{-E_p(t+t'-2t_0)} + T\, \frac{\mathrm{e}^{-E_p|t-t'|} - \mathrm{e}^{-E_p(t+t'-2t_0)}}{E_p} \end{aligned} \tag{3.10}$$

As $t_0 \to -\infty$, we approach the 'equilibrium limit'

$$C_0(p;t,t') = \frac{T\mathrm{e}^{-E_p|t-t'|}}{E_p}. \tag{3.11}$$

The correlation now only depends upon the time difference (time translational invariance or TTI). Its (time) Fourier Transform is then

$$C_0(p;\omega) = \frac{T}{E_p} \int_{-\infty}^{+\infty} \mathrm{d}\tau\, \mathrm{e}^{+\mathrm{i}\omega\tau} \mathrm{e}^{-E_p|\tau|} = \frac{2T}{|-\mathrm{i}\omega + E_p|^2}. \tag{3.12}$$

- *Response*
We note that in Eq. (3.1) the noise $\eta(x;t)$ is homogeneous to an external time dependent source (or magnetic field), so that the response function can equivalently be computed by deriving the average field with respect to the noise. Thus we have

$$R_0(p;t,t') = \left\langle \frac{\delta\phi(p;t)}{\delta\eta(p;t')} \right\rangle = \mathrm{e}^{-E_p(t-t')}\Theta(t-t'). \tag{3.13}$$

The response is causal (i.e. $R_0 = 0$ if $t < t'$). At *this level* R_0 is only $t - t'$ dependent, and

$$R_0(p;\omega) = \frac{1}{-\mathrm{i}\omega + E_p}. \tag{3.14}$$

Note that a *fluctuation–dissipation* relationship holds in the equilibrium limit, relating R_0 and C_0:

$$\frac{1}{T}\frac{\mathrm{d}}{\mathrm{d}\tau} C_0(p;\tau) = -R_0(p;\tau), \quad \tau > 0, \tag{3.15}$$

or

$$\frac{1}{T}\,\omega\, C_0(p;\omega) = 2\,\mathrm{Im}\, R_0(p;\omega), \tag{3.16}$$

or, after integration,

$$\frac{1}{T}[C_0(p;\tau=0) - C_0(p;\tau)] = \int_0^\tau R(p;\tau')\,\mathrm{d}\tau'. \tag{3.17}$$

This relation, also known as the dynamical Fluctuation–Dissipation Theorem (FDT), is a very important and quite generic feature of equilibrium dynamics. It can be proved under very general circumstances also for the complete correlation and response function. A proof order by order can be found in Ma (1976); another very general proof is given by Kurchan (1992), where the dynamical functional is written in terms of a supersymmetric effective Lagrangian (much as Parisi and Sourlas did for the static tree expansion, see Section 2.4.1): in this case the dynamical Fluctuation–Dissipation Theorem naturally arises as a Ward–Takahashi Identity generated by the supersymmetry. When disorder is present, it may, however, happen that equilibrium is never truly attained, the average correlation and response function do not satisfy time translation invariance and the fluctuation–dissipation relation does not hold.

3.1.2 Perturbation expansion

Let us now proceed to implement a perturbative expansion around the bare average values. To better see the structure of the expansion, we simplify the notation, omitting the momenta and writing explicitly only the time dependence. Also, it is convenient to set $\phi_0 = 0$. Then, using Expression (3.13), Eq. (3.6) becomes

$$\phi(t) = \int_{t_0}^{t} \mathrm{d}s\, R_0(t-s)\left[\eta(s) - \frac{u_1}{6}\phi^3(s)\right]. \tag{3.18}$$

Expanding in powers of u_1, we can build the representative graphs with the following convention:

$R_0(t-s)$: t •←• s

$\eta(s)$: —✱

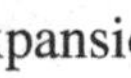

vertex: ✱

With this notation the series expansion for the field $\phi(t;t_0)$ can be formally written as:

$$\phi(t) = \underset{t}{\bullet}\!\!-\!\!<\!\!-\!\!\ast \;+\; \frac{1}{3!}\,[\text{graph}] \;+\; \frac{1}{2\cdot 3!}\,[\text{graph}] \;+\; \cdots, \tag{3.19}$$

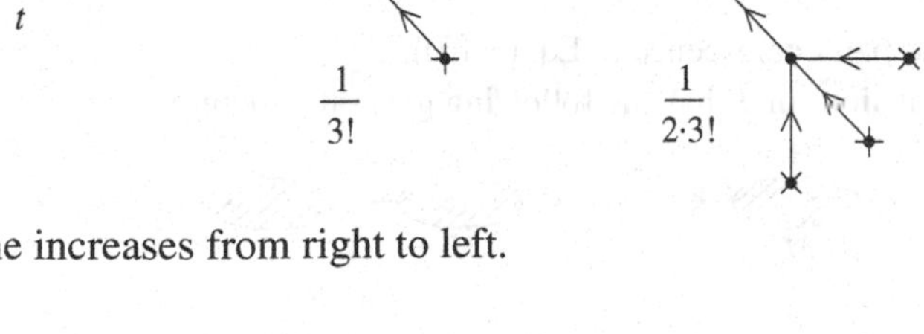

where time increases from right to left.

Using this formalism, one can easily obtain the graphical expansion for the average quantities.

- *Magnetization*
  Let us consider first the one-point average. Since the series is *odd* in η in the absence of a magnetic field we have $\langle\phi(t)\rangle = 0$. In the presence of a field H, $\eta(s)$ would be replaced by $\eta(s) + H$, and

$$\langle\phi(t)\rangle = \frac{1 - e^{-E_p(t-t_0)}}{E_p} H + \cdots , \tag{3.20}$$

  and if $t_0 \to -\infty$ (equilibrium limit):

$$\langle\phi\rangle = \frac{H}{E_p}. \tag{3.21}$$

- *Response*
  The response is defined by $R = \langle\partial\phi(t)/\partial\eta(t')\rangle$. From the above series (3.19) we get a (single) tree expansion for the field derivative:

$$\frac{\partial\phi(t)}{\partial\eta(t')} = \cdots + \frac{1}{2}\cdots + \frac{1}{4}\cdots + \cdots \tag{3.22}$$

Averaging over the thermal noise η (the series is now *even* in η) we finally obtain

$$R(t,t') \equiv \left\langle \frac{\partial\phi(t)}{\partial\eta(t')} \right\rangle = \cdots + \cdots + \frac{1}{4}\cdots + \frac{1}{2}\cdots + \cdots , \tag{3.23}$$

where we have coalesced pairs of ηs and we have used the notation

$$C_0(t,s) \equiv \langle \cdots \rangle_\eta \equiv \bullet\!-\!\otimes\!-\!\bullet$$

(as it results from the pair coalescence in Eq. (3.10)).

Note that the expansion for R has the following general structure:

$$R = \cdots + \cdots M \cdots + \cdots M \cdots M \cdots + \cdots , \tag{3.24}$$

where M is the 'mass operator' i.e. the sum of all one-particle irreducible diagrams:

$$M(t,t') = \quad + \quad + \quad \cdots \tag{3.25}$$

This results in the Dyson equation:

$$R = R_0 - R_0 M R, \qquad \text{or} \qquad R^{-1} = R_0^{-1} + M. \tag{3.26}$$

- *Correlation*
The perturbative expansion for the correlation function $C = \langle \phi(t)\phi(t') \rangle$ is given by a double tree, i.e., for example,

$$\left(\quad t \quad s_1 \quad s_2 \quad s_3 \quad \right) \left(\quad s_1' \quad s_4' \quad s_2' \quad s_3' \quad s_5' \quad t' \quad \right). \tag{3.27}$$

Note that no ordering between the times of (s, t) and of (s', t') is required in this case.

The general structure after η-averaging is easily inferred:

$$C(t,t') = R(t;s)\hat{M}(s;s')R(s';t'), \tag{3.28}$$

where the new mass operator

$$\hat{M}(s;s') = \quad + \quad + \quad \cdots \tag{3.29}$$

is *one-particle irreducible* again (with respect to cutting one R_0 or one C_0 line) but such that if one dissolves all the pairs of η coalescence ($\otimes$) it *falls into two disconnected pieces* (and *only* two of them). For example:

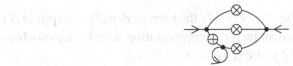

this graph is not permitted as a contribution to $\hat{M}$ since it falls into *three* disconnected pieces when disconnecting all the bare correlators.

3.2 Martin–Siggia–Rose formulation

The previous perturbative expansion is standard in dynamical field theory (see, e.g., Ma, 1976), but is not very convenient when dealing with systems with quenched disorder. An alternative formulation, that can be easily generalized to the disordered case, is the so-called *generating functional* approach (Martin, Siggia and Rose,

1973). It is based on the following simple identity:

$$1 = \int \prod_i \mathrm{D}\phi_i(t)\delta\left[\frac{\partial \phi_i(t)}{\partial t} + \frac{\partial \mathcal{H}}{\partial \phi_i(t)} - \eta_i(t)\right] \times \left|\det\left(\frac{\partial}{\partial t}\delta(t-t')\delta_{i;j} + \frac{\partial^2 \mathcal{H}}{\partial \phi_i(t)\partial \phi_j(t)}\delta(t-t')\right)\right|, \tag{3.30}$$

where $\mathrm{D}\phi_i(t)$ indicates functional integration over the function $\phi_i(t)$. In this expression we have δ functions on the dynamical equation (for each value of i) with the appropriate Jacobian $\det(\mathcal{J}_{ij})$, integrated over their arguments.

Equation (3.30) can be further explicited if we use an integral representation for the δ functions:

$$\delta\left(\mathcal{F}[\phi(t)]\right) = \int \mathrm{D}\hat{\phi}(t) \exp\left(\int \mathrm{d}t\, \mathrm{i}\hat{\phi}(t)\mathcal{F}[\phi(t)]\right), \tag{3.31}$$

where we include for convenience the normalization factor in the integration measure. We also note that, for a given initial condition and thermal noise, the Langevin equation admits a single causal solution. Thus *the Jacobian does not change sign* and one can safely remove the absolute value sign. One can thus build a generating functional for the dynamics as

$$\hat{Z}\{\ell_i, \hat{\ell}_i\} \equiv \left\langle \int \prod_i \left(\mathrm{D}\phi_i(t)\mathrm{D}\hat{\phi}_i(t)\right)\, \mathrm{e}^{\sum_i \int \mathrm{d}t[\ell_i(t)\phi_i(t) + \hat{\ell}_i(t)\hat{\phi}_i(t)]}\, \mathrm{e}^{\mathcal{L}_\eta\{\hat{\phi}_i, \phi_i\}} \right\rangle, \tag{3.32}$$

where ℓ_i and $\hat{\ell}_i$ are two source fields and

$$\mathcal{L}_\eta\{\hat{\phi}_i, \phi_i\} = \sum_i \int \mathrm{d}t\, \mathrm{i}\hat{\phi}_i(t)\left[\frac{\partial}{\partial t}\phi_i(t) + \frac{\partial \mathcal{H}}{\partial \phi_i(t)} - \eta_i(t)\right] + \mathrm{tr}\ln\left\{\frac{\partial}{\partial t}\delta(t-t')\delta_{i;j} + \frac{\partial^2 \mathcal{H}}{\partial \phi_i(t)\,\partial \phi_j(t)}\delta(t-t')\right\}. \tag{3.33}$$

Here we have used for the Jacobian $\mathcal{J}_{ij} = \partial_i \partial_j \mathcal{H}$ the identity $\det(\mathcal{J}) = \exp(\mathrm{tr}\ln \mathcal{J})$. Alternatively, we could have used an integral representation with Fermi variables: $\det \mathcal{J} = \int D\, c_i(t) D\, \bar{c}_i(t)\, \exp\{\sum \int \mathrm{d}t\; c_i(t)\mathcal{J}_{ij}(t,t')\, \bar{c}_j(t,t')\}$.

We can now perform explicitly the *noise-average* over the probability law (3.2) to get

$$\hat{Z}\{\ell_i, \hat{\ell}_i\} \equiv \int \prod_i \left(\mathrm{D}\phi_i(t)\mathrm{D}\hat{\phi}_i(t)\right)\, \mathrm{e}^{\sum_i \int \mathrm{d}t[\ell_i(t)\phi_i(t) + \hat{\ell}_i(t)\hat{\phi}_i(t)]}\, \mathrm{e}^{\mathcal{L}\{\hat{\phi}_i, \phi_i\}}, \tag{3.34}$$

where now the effective Lagrangian $\mathcal{L}$ is given by

$$\mathcal{L}\{\hat{\phi}_i, \phi_i\} = \sum_i \int \mathrm{d}t \left\{ \mathrm{i}\hat{\phi}_i(t)\left[\frac{\partial}{\partial t}\phi_i(t) + \frac{\partial \mathcal{H}}{\partial \phi_i(t)}\right] - T|\hat{\phi}_i(t)|^2 \right\} + \mathrm{tr}\ln \mathcal{J}_{ij}(t;t'). \tag{3.35}$$

The functional $\hat{Z}$ can be used to generate all the average correlation and response functions associated with the dynamical process described by the Langevin equation (3.1). For example, we have, by construction,

$$C(t,t') = \frac{1}{N}\sum_i \left.\frac{\partial^2 \hat{Z}\{\ell_i, \hat{\ell}_i\}}{\partial \ell_i(t)\partial \ell_i(t')}\right|_{\ell,\hat{\ell}=0} . \tag{3.36}$$

Also, we note that if an external magnetic field H is considered, the source $\hat{\ell}$ is homogeneous to $-\mathrm{i}H$, therefore we have

$$R(t,t') = -\frac{\mathrm{i}}{N}\sum_i \left.\frac{\partial^2 \hat{Z}\{\ell_i, \hat{\ell}_i\}}{\partial \ell_i(t)\partial \hat{\ell}_i(t')}\right|_{\ell,\hat{\ell}=0} . \tag{3.37}$$

The correlation and response functions can be expressed through Eqs. (3.36) and (3.37) in terms of a functional measure in the space of the fields ϕ and $\hat{\phi}$, e.g.:

$$\begin{aligned} C_i(t,t') &= \langle \phi_i(t)\phi_i(t')\rangle = \langle \phi_i(t)\phi_i(t')\rangle_{\mathcal{L}}, \\ R_i(t,t') &= \left\langle \frac{\partial \phi_i(t)}{\partial H(t')}\right\rangle = -\mathrm{i}\langle \phi_i(t)\hat{\phi}_i(t')\rangle_{\mathcal{L}}, \end{aligned} \tag{3.38}$$

where at the r.h.s. the angle brackets indicate averages over the functional measure defined in (3.32). In the following we shall often omit the subscript $\mathcal{L}$ when there is no ambiguity.

As can be seen from the previous expressions, the functional $\hat{Z}$ has much the same role as the partition function in the statics. Note, however, that, by construction, $\hat{Z}\{\ell = 0\} = 1$, and therefore when evaluating averages with the measure defined in (3.32) one does not need to normalize. This fact has important consequences when dealing with the disordered case: in the statics it was precisely the explicit dependence of the partition function on the quenched disorder that required the introduction of replicas. In the dynamics, on the other hand, there is *no need to introduce replicas, since the norm is not randomness dependent.* This will be used later when going back to the RFIM.

3.2.1 Perturbation expansion from MSR

The MSR formalism provides us with an effective dynamical Lagrangian, analogous to the Lagrangian occurring in the statics. Much in the same way, it is then possible to develop a perturbative expansion in the coupling u_1.

Since for the moment we are only interested in the equilibrium limit, we may send the initial time to $-\infty$, and *benefit from time-translational invariance.* Expanding in the coupling u_1, we then have to average products of $\hat{\phi}$, ϕ with the bare Lagrangian:

$$\mathcal{L}^0 = \sum_p \int \frac{\mathrm{d}\omega}{2\pi}\{\mathrm{i}\hat{\phi}_p(\omega)[-\mathrm{i}\omega + E_p]\phi_{-p}(-\omega) - T|\hat{\phi}_p(\omega)|^2\}. \tag{3.39}$$

The bare correlation and response functions of Section 3.1.2 then appear in this context as *bare propagators* of the effective dynamical field theory, that is, in the *equilibrium limit*,

$$\langle \phi_p(\omega)\phi_{-p}(-\omega)\rangle_0 = C_0(p;\omega) = \frac{2T}{|-\mathrm{i}\omega + E_p|^2}, \tag{3.40}$$

$$-\mathrm{i}\langle \phi_p(\omega)\hat{\phi}_{-p}(-\omega)\rangle_0 = R_0(p;\omega) = \frac{1}{-\mathrm{i}\omega + E_p}. \tag{3.41}$$

And, by causality,

$$\langle \hat{\phi}_p(\omega)\hat{\phi}_{-p}(-\omega)\rangle_0 = 0. \tag{3.42}$$

Note that the noise average has automatically been taken with the effective Lagrangian, and we can write *directly* the contributions to R, C and higher order correlation functions. This can be done more easily by defining a graphical representation for the fields. We shall use straight lines to represent the original fields ϕ and wavy lines for the conjugated ones $\hat{\phi}$. In this way:

$C_0(t,t')$:

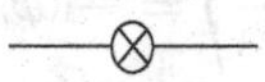

$R_0(t,t')$:

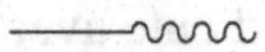

$\frac{u_1}{3!}\phi^3\mathrm{i}\hat{\phi}$:

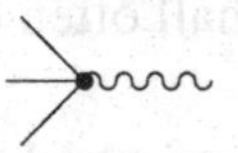

Note the new graphical representation for the bare response function, that replaces the one used in the tree expansion of the previous section.

The perturbative expansion of the correlation and response function can be easily expressed in terms of the corresponding mass operators that, as usual, include all the one-particle irreducible diagrams. The Dyson equations are formally the same as in (3.26) and (3.28), with the mass operators now expressed in the new graphical representation. If we are interested in the *equilibrium limit* we can assume TTI and the Dyson equations are diagonalized via a Fourier Transform, therefore we have

$$R(p,\omega) = [-\mathrm{i}\omega + E_p + M(p,\omega)]^{-1} \tag{3.43}$$

with

$$M(p,\omega) = \quad + \quad + \cdots \tag{3.44}$$

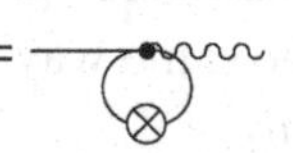

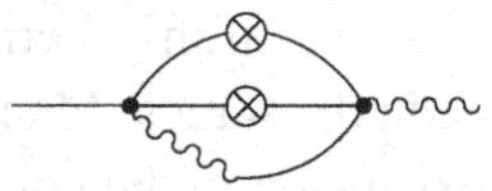

$\frac{1}{2}$ $\frac{1}{2}$

Likewise for the correlation,

$$C(p,\omega) = R(p,\omega)\,\hat{M}(p,\omega)\,R^*(p,\omega) \tag{3.45}$$

with

$$\hat{M}(p,\omega) = \quad + \quad + \cdots \tag{3.46}$$

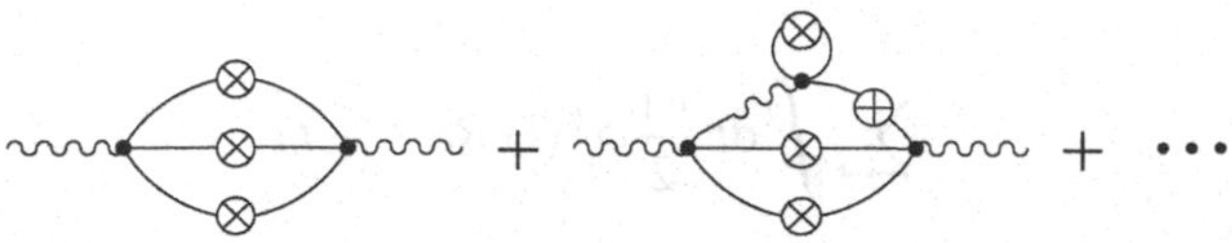

$$\frac{1}{2}\cdot\frac{1}{3!} \qquad\qquad \frac{1}{2}$$

Similarly, for the 4-point vertex functions we find

$$\frac{1}{6}\Gamma^{(4)}_{\hat{\phi}\phi^3} = \quad + \quad + \cdots \tag{3.47}$$

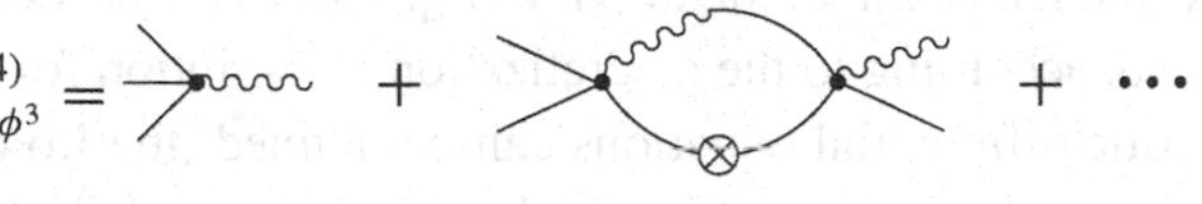

$$\frac{1}{6} \qquad\qquad \frac{1}{2}$$

and

$$\frac{1}{4}\Gamma^{(4)}_{\hat{\phi}^2\phi^2} = \quad + \cdots \tag{3.48}$$

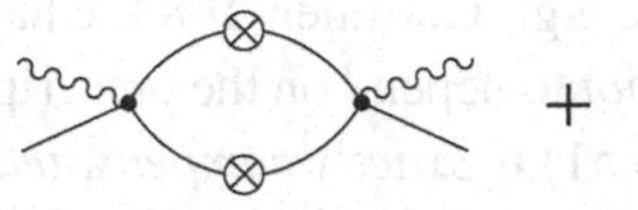

$$\frac{1}{4}$$

If we take p and ω to be the *total* momentum and frequency entering the 4-point graph, then (3.47) becomes

$$\frac{1}{6}\Gamma^{(4)}_{\hat{\phi}\phi^3}(p;\omega) = \frac{u_1}{6} - \frac{u_1^2}{2}\int \mathrm{d}\Omega \int \frac{\mathrm{d}^D q}{(2\pi)^D}\frac{1}{(-\mathrm{i}\Omega + E_q)}\frac{2T}{|-\mathrm{i}\omega + \mathrm{i}\Omega + E_{p-q}|^2} + \cdots \tag{3.49}$$

The graph expansion so described remains *unchanged* for $t_0 \neq -\infty$ if one takes $\phi_0(p) = 0$ as initial condition. However, $C_0\left(p;t,t'\right)$ has in this case terms depending on t_0 (see Eq. (3.10)) and the Fourier Transform to frequencies is not very useful. The bare response $R_0(p;t,t')$ is unchanged as in (3.13).

3.2.2 Role of the Jacobian

So far we have neglected the term

$$\operatorname{tr}\ln\left\{\left[\frac{\partial}{\partial t}\delta_{i;j} + \frac{\partial^2 H(\phi)}{\partial\phi_i\partial\phi_j}\right]\delta(t-t')\right\} \tag{3.50}$$

in $\mathcal{L}$. Inside the ln, the bare term is nothing but $[R_0^{-1}](t,t')$, i.e. the inverse of the causal response (bare propagator). The coupling term is $\frac{u_1}{2}\phi_i^2(t)$.

Expanding in u_1, we get closed loops since we have a trace, built with R_0 propagators. Each one of these propagators, we recall, brings a Θ function with respect to its arguments, expressing the causal nature of the response. Hence we get zero contributions (one cannot close loops of Θ-function propagators) except for the term

$$\sum_i \int \mathrm{d}t \, \frac{u_1}{2} \phi_i^2 (t) \, R_0 (t - t) \,. \tag{3.51}$$

The Θ function of zero argument is not in general well defined. In the context of Langevin dynamics, the value of the response function at zero argument actually depends on the particular interpretation of the stochastic equation one is considering. From a mathematical point of view Eq. (3.1) can be defined only in an integral form and, according to the discretization prescription for the integral, two possible stochastic differential equations can be defined: the Ito equation and the Stratonovich one. In the first case, the field depends only on the noise at previous times, the response $R(0)$ is therefore zero at all orders (while $\lim_{t \to 0^+} R(t) = 1$) and the Jacobian does not give any finite contribution. In the Stratonovich case this is, however, not true (see, e.g., Gardiner, 1983, Chapter 3). Obviously we expect the physical observables *not* to depend on the prescription adopted. Indeed, one can easily see that the term (3.51) is *exactly compensated* by all terms of the perturbation expansion where, in expanding in $(u_1/6) \sum \int \hat{\phi}_i(t) \phi_i^3(t) \, \mathrm{d}t$, one contracts the $\hat{\phi}(t)$ with one of the $\phi(t)$ of the *same* vertex.

As a result, whatever prescription is adopted for the Langevin equation, one may keep the same perturbation expansion as described in the previous section, but *omitting all contributions containing the insertion:*

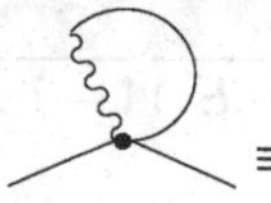

$$\equiv -\frac{u_1}{2} R_0(t - t' = 0). \tag{3.52}$$

$\frac{1}{2}$

3.2.3 Fluctuation–Dissipation Theorem

As anticipated before, in the equilibrium limit the Fluctuation–Dissipation Theorem (FDT) can be generalized to all orders, thus we have

$$\frac{1}{2T} \, \omega \, C \, (\omega) = \mathrm{Im} R \, (\omega) \,. \tag{3.53}$$

Since $R\,(\omega)$ is an analytic function, we have (Kramers–Krönig relationship)

$$\mathrm{Re} R \, (\omega) = \int_{-\infty}^{+\infty} \frac{\mathrm{d}\omega'}{\pi} \frac{\mathrm{Im} R(\omega')}{\omega' - \omega} , \tag{3.54}$$

where integrals are meant in principal part. Hence,

$$\int_{-\infty}^{+\infty} \frac{\mathrm{d}\omega}{2\pi} C \, (\omega) = T \int_{-\infty}^{+\infty} \frac{\mathrm{d}\omega'}{\pi} \frac{\mathrm{Im} R(\omega')}{\omega'} = T \, R(\omega = 0), \tag{3.55}$$

i.e. the equal-time correlation function is T times the zero-frequency limit of the response.

3.2.4 Static limit

As we said, we have taken the equilibrium limit by sending the initial time t_0 to minus infinity. We then expect one-time quantities to approach the corresponding static values, i.e. the results obtained working directly with the static partition function

$$Z = \int \mathrm{D}\phi \, \mathrm{e}^{-\beta\mathcal{H}(\phi)}.$$

Let us now discuss this point for correlation and response functions.

- *Two-point functions*

  The static limit of the equal-time correlation function is immediately given by Eq. (3.55):

$$C_{st} = \lim_{t_0 \to -\infty} \langle \phi(t)\, \phi(t) \rangle = \int_{-\infty}^{+\infty} \frac{\mathrm{d}\omega}{2\pi} C(\omega) = T R(\omega = 0)\,. \tag{3.56}$$

  This can be shown order by order to give precisely the static value $C = \langle \phi\phi \rangle$, where *the averages are now taken with respect to the static measure*. Let us verify it *to one loop,* and to make things lighter let us consider a coupling $(u_1/6)\, \phi_i^3$, instead of $(u_1/24)\, \phi_i^4$. To *one loop* we have

$$R(\omega) = -\mathrm{i} < \phi(\omega)\hat{\phi}(-\omega) >= \text{[diagram: lines } E,\ E \text{; loop } E_1,\ E_2 \text{]}$$

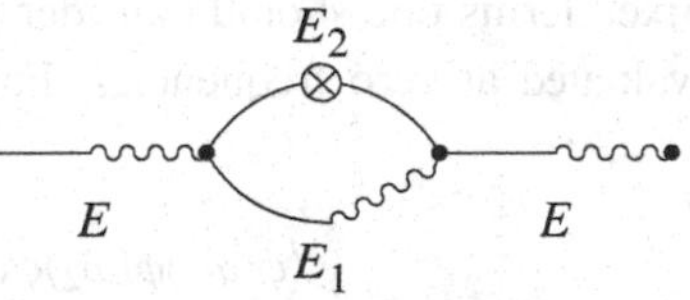

$$= u_1^2 \sum_{\substack{\Omega\\1,2}} \left\{ \frac{1}{-\mathrm{i}\omega + E} \frac{1}{-\mathrm{i}\omega + \mathrm{i}\Omega + E_1} \frac{2T}{|-\mathrm{i}\Omega + E_2|^2} \frac{1}{-\mathrm{i}\omega + E} \right\}, \tag{3.57}$$

  where the sum indicates an integral over the frequency Ω and over the momenta p_1 and p_2. From this we get

$$\begin{aligned} T R(\omega = 0) &= T \frac{u_1^2}{E^2} \sum_{\substack{\Omega\\1,2}} \frac{1}{-\mathrm{i}\Omega + E_1} \frac{2T}{|-\mathrm{i}\Omega + E_2|^2} = T \frac{u_1^2}{E^2} \sum_{1,2} \frac{1}{E_1 + E_2} \frac{2T}{2E_2} \\ &= \frac{T}{2} \frac{u_1^2}{E^2} \sum_{1,2} \frac{2T}{E_1 + E_2} \left[\frac{1}{2E_1} + \frac{1}{2E_2} \right] \\ &= \frac{T^2}{2} \frac{u_1^2}{E^2} \sum_{1,2} \frac{1}{E_1 E_2}, \end{aligned} \tag{3.58}$$

which is indeed the result that one would have obtained directly from the static expansion:

$$C = \langle \phi\phi \rangle =$$

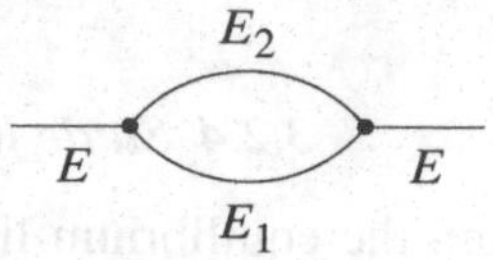

$$= \frac{T^2}{2}\frac{u_1^2}{E^2}\sum_{1,2}\frac{1}{E_1 E_2}. \tag{3.59}$$

This result also illustrates what is the correct static limit to be taken on the response function. Indeed, from the static Fluctuation–Dissipation Theorem we have that $\langle \phi\phi \rangle = T\partial\langle\phi\rangle/\partial H$, which immediately gives

$$R(\omega = 0) = \frac{\partial\langle\phi\rangle}{\partial H}. \tag{3.60}$$

This relation tells us the rather intuitive fact that, when the system equilibrates, the integrated dynamical response function $\int_0^t \mathrm{d}t'\, R(t')$ approaches the static response, i.e. the susceptibility. Note also that Eq. (3.55) already encodes the static fluctuation–dissipation relation.

- *Multipoint functions*
For correlations at equilibrium, again equal-time functions $\langle \phi(t)\ldots\phi(t)\rangle$ give the correct static limit (Parisi, unpublished). For pure responses, i.e. functions of the kind $\langle \phi(\omega_1)\hat{\phi}(\omega_2)\ldots\hat{\phi}(\omega_n)\rangle$ one can prove that, as for the two-point response, zero frequencies give the correct static limit (De Dominicis, 1975). In the same way, for more complicated mixed terms one should consider the fields ϕ evaluated at equal times and the fields $\hat{\phi}$ evaluated at zero frequencies. For example, the static limit of the four-point function

$$\langle \phi(\omega_1)\phi(\omega_2)\hat{\phi}(\omega_3)\hat{\phi}(\omega_4)\rangle \delta^{Kr}_{\sum\omega;0} \tag{3.61}$$

is given by

$$I_{st}^{(2)} = \int_{-\infty}^{-\infty} \frac{\mathrm{d}\omega}{2\pi} \langle \phi(\omega)\phi(-\omega)\hat{\phi}(0)\hat{\phi}(0)\rangle, \tag{3.62}$$

where we have taken equal times for the first two fields, and zero frequency for the last two. The other possible four-point functions involve all the remaining combinations among the fields ϕ and $\hat{\phi}$. When evaluating their static limit one would have to consider

$$\begin{aligned}
I_{st}^{(1)} &= \int_{-\infty}^{-\infty} \frac{\mathrm{d}\omega_1}{2\pi}\frac{\mathrm{d}\omega_2}{2\pi} \langle \phi(\omega_1)\phi(\omega_2)\phi(-\omega_1-\omega_2)\hat{\phi}(0)\rangle, \\
I_{st}^{(0)} &= \int_{-\infty}^{+\infty} \frac{\mathrm{d}\omega_1}{2\pi}\frac{\mathrm{d}\omega_2}{2\pi}\frac{\mathrm{d}\omega_3}{2\pi} \langle \phi(\omega_1)\phi(\omega_2)\phi(\omega_3)\phi(-\omega_1-\omega_2-\omega_3)\rangle, \\
I_{st}^{(3)} &= \langle \phi(0)\hat{\phi}(0)\hat{\phi}(0)\hat{\phi}(0)\rangle.
\end{aligned} \tag{3.63}$$

It turns out that, by means of fluctuation–dissipation relations, *all these* expressions are equivalent, i.e. $I_{st}^{(p)} = I_{st}^{(0)}(T)^p$ (except $\langle\hat{\phi}\ldots\hat{\phi}\rangle = 0$ by causality). From a practical point of view then, if we know some dynamical equilibrium result, the static limit is most easily obtained by sticking to pure responses $\langle\phi\hat{\phi}\ldots\hat{\phi}\rangle$ and *setting all external frequencies to zero.*

3.3 *RFIM* dynamics

Now we want to apply the dynamical approach to the RFIM (Brezin and De Dominicis, 1999). We observed in Section 3.2 that the norm of $\hat{Z}$ is equal to one. This means that, when quenched disorder is present, we are allowed to perform the average over the disorder directly on $\hat{Z}$. Thus, the dynamical functional used to compute physical observables becomes

$$\overline{\hat{Z}}\{\ell_i, \hat{\ell}_i\} \equiv \overline{\left\langle \int \prod_i \left(\mathrm{D}\phi_i(t)\mathrm{D}\hat{\phi}_i(t)\right) \, \mathrm{e}^{\sum_i \int \mathrm{d}t [\ell_i(t)\phi_i(t) + \hat{\ell}_i(t)\hat{\phi}_i(t)]} \, \mathrm{e}^{\mathcal{L}_{\eta;h}\{\hat{\phi}_i, \phi_i\}} \right\rangle}, \quad (3.64)$$

where, as usual, the bar indicates averages over the quenched disorder. Consequently, the effective measure and the effective Lagrangian $\mathcal{L}$ obtained from (3.64) by explicitly averaging over η and h (the analogue of (3.34)) encode both the averages, the thermal and the quenched ones. The same is true for correlation and response functions and, instead of (3.38), we have

$$C_i(t, t') = \overline{\langle\phi_i(t)\phi_i(t')\rangle} = \langle\phi_i(t)\phi_i(t')\rangle_{\mathcal{L}},$$
$$R_i(t, t') = \overline{\left\langle \frac{\partial\phi_i(t)}{\partial h(t')} \right\rangle} = -\mathrm{i}\langle\phi_i(t)\hat{\phi}_i(t')\rangle_{\mathcal{L}}. \quad (3.65)$$

Let us now look at the case of the RFIM. We consider as a starting point the ϕ^4 Hamiltonian appearing in (2.7), and the Langevin equation associated with it. That is, we simply add a random field h_i to a pure ϕ^4 model and we do not consider at this level the generalized couplings of the previous chapter. We can then easily generalize the MSR approach described so far for the pure system. Since the random field h_i is *time independent*, the bare quadratic Lagrangian, after η- and h-averaging, acquires an extra Δ-term (see Eq. (3.39))

$$\mathcal{L}^0_{\mathrm{RFIM}} = \int \frac{\mathrm{d}^D p}{(2\pi)^D} \int \frac{\mathrm{d}\omega}{2\pi} \{\mathrm{i}\hat{\phi}_p(\omega)[-\mathrm{i}\omega + E_p]\phi_p(\omega) - T|\hat{\phi}_p(\omega)|^2\}$$
$$+\Delta \int \frac{\mathrm{d}^D p}{(2\pi)^D} \left| \int \mathrm{d}\omega \delta(\omega) \hat{\phi}(p;\omega) \right|^2 . \quad (3.66)$$

From this expression we immediately see that the bare correlation function is changed from the pure case because of the Δ-term, indeed we get

$$\langle \phi_p(\omega)\,\phi_{-p}(-\omega)\rangle_0 = \left[\frac{2T}{\left|-\mathrm{i}\omega + E_p\right|^2}\right] + \left[2\pi\,\delta(\omega)\,\frac{\Delta}{E_p^2}\right]$$

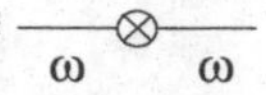

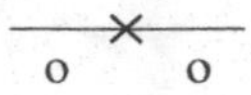

$$\equiv \underset{\omega\qquad\omega}{\text{—}\otimes\text{—}} + \underset{0\qquad 0}{\text{—}\times\text{—}} \tag{3.67}$$

The first term in (3.67), the one present also in the pure case, comes from the thermal noise average and we shall refer to it as the T-line. The second term comes from the random field average and we have introduced for it a corresponding graphical representation that we shall call the Δ-line. Note that this (bare) correlation exhibits a component in $\delta(\omega)$. This is a new feature which persists also in the full correlation. The bare response and the coupling term are the same as before. Thus, we have

$$-\mathrm{i}\big\langle \phi_p(\omega)\,\hat{\phi}_{-p}(-\omega)\big\rangle_0 = \frac{1}{-\mathrm{i}\omega + E_p} \equiv \underset{\omega}{\text{—}\sim\sim} \tag{3.68}$$

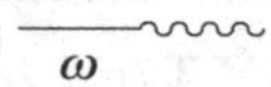

and for the *vertex*

$$\frac{u_1}{6}\phi_{p_1}(\omega_1)\,\phi_{p_2}(\omega_2)\,\phi_{p_3}(\omega_3)\,\hat{\phi}_{p_4}(\omega_4)\,\delta^{Kr}_{(\sum_i p_i;0)}\,\delta\left(\sum_i \omega_i\right) \equiv \succ\!\!\bullet\!\sim\sim \tag{3.69}$$

Note that we could also have applied the same procedure as in Section 3.1.2 and calculated the $\langle\phi\phi\rangle$ correlation directly from Eq. (3.6), where one would have inserted a random *field* replacing $\eta(p;s)$ by $\eta(p;s) + h(p)$. This would have resulted in, instead of (3.10):

$$C_0(p;t,t') = \phi_0^2(p)\,\mathrm{e}^{-E_p(t+t'-2t_0)} + T\,\frac{\mathrm{e}^{-E_p|t-t'|} - \mathrm{e}^{-E_p(t-t'-2t_0)}}{E_p}$$
$$+\,\Delta\left[\frac{1-\mathrm{e}^{-E_p(t-t_0)}}{E_p}\right]\left[\frac{1-\mathrm{e}^{-E_p(t'-t_0)}}{E_p}\right]. \tag{3.70}$$

The last term, under $t_0 \to -\infty$, becomes in Fourier Transform $\Delta\delta(\omega)/E_p^2$ i.e. the Δ-line contribution.

Let us examine what are the consequences of the new Δ-line on the perturbative expansion of the basic two- and four-point functions:

- *Response*

  In each term of (3.44) each T-line must be replaced by T-line + Δ-line. In the absence of randomness ($\Delta = 0$), i.e. for the pure system, we know that the double denominators of the T-lines are reduced to the single ones of the static limit through frequency integration (see Eq. (3.58)). In this same limit the new, zero-frequency, double denominators of the Δ-lines are there to stay and generate the static infrared divergences discussed in the previous chapter. As we did in the statics within the replica approach to investigate the infrared behaviour, we then need to consider the *maximum* of Δ-lines, corresponding to the most divergent graphs. We have in this way

$M\,(p;\omega) = \cdots +$ 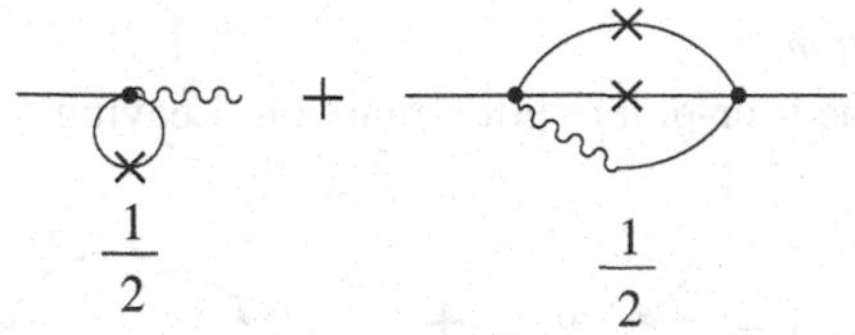 , (3.71)

where we have omitted terms that do not have the maximum number of Δ-lines.

- *Correlation*

Again, we need to replace all T-lines with $T + \Delta$-lines in (3.46). If *all the T-lines* that connect the left and right pieces together are *not* changed into Δ-lines, we get contributions to $\hat{M}\,(p;\omega)$ that remain subdominant. If *all* of them *become Δ-lines*, then we get a contribution in $\delta\,(\omega)$! It is then convenient to explicitly separate the zero-frequency mode and write the correlation as

$$\begin{aligned} C\,(p;\omega) &= C_T\,(p;\omega) + 2\pi\; C_\Delta\,(p)\,\delta\,(\omega) \\ &= R\,(p;\omega)\,\hat{M}_T\,(p;\omega)\,R^*\,(p;\omega) + R\,(p;0)\left[2\pi\delta\,(\omega)\,\hat{M}_\Delta\,(p)\right]R\,(p;0)\,, \end{aligned} \tag{3.72}$$

with

$\hat{M}_T\,(p;\omega) =$

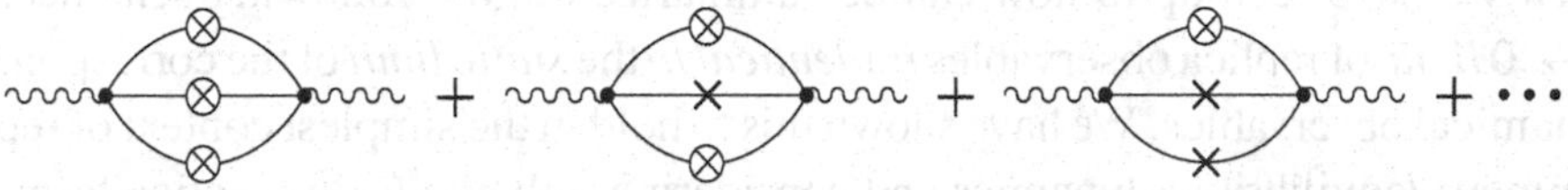 (3.73)

and

$\delta\,(\omega)\,\hat{M}_\Delta\,(p) =$ 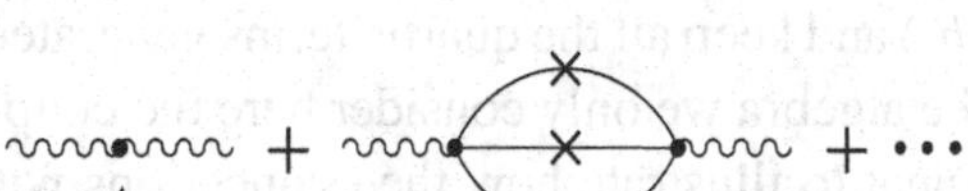 (3.74)

Note that $\hat{M}_\Delta\,(p)$ is the term that becomes disconnected under undoing the h-average.

This expression has to be compared with the static propagator of the replica calculation, that is, in absence of replica symmetry breaking,

$$G^{ab}\,(p) \equiv G_a\,(p)\,\delta_{a;b} + C_{ab}\,(p) \equiv G\,(p)\,\delta_{a;b} + C\,(p), \tag{3.75}$$

where, we remind, $C(p)$ represents the h-connected contribution. Therefore, in the *static* limit we have the correspondence

$$\int \frac{\mathrm{d}\omega}{2\pi} C_T\,(p;\omega) \Longrightarrow G_a\,(p) = G\,(p) \tag{3.76}$$

and

$$\int \mathrm{d}\omega\, \delta(\omega)\, C_\Delta(p) \Longrightarrow C_{ab}(p) = C(p). \tag{3.77}$$

- *Four-point function*
  Let us look at the four-point vertex function. Leaving apart the subdominant terms, we have

$$\frac{1}{6}\Gamma^{(4)}_{\hat{\phi}\phi^3}(\omega) = \quad + \quad + \cdots \tag{3.78}$$

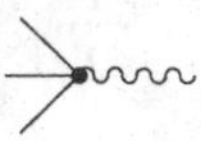

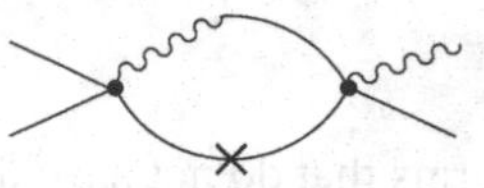

$$\frac{1}{6} \qquad\qquad \frac{1}{2}$$

That is,

$$\frac{1}{6}\Gamma^{(4)}_{\hat{\phi}\phi^3}(\omega) = \frac{u_1}{6} - \frac{u_1^2}{2}\int \frac{\mathrm{d}q^D}{(2\pi)^D}\frac{1}{-\mathrm{i}\omega + E_q}\frac{\Delta}{E^2_{p-q}} \tag{3.79}$$

which in the static limit ($\omega = 0$) gives back (2.22):

$$\Gamma^{(4)}_{\hat{\phi}\phi^3}(\omega = 0) \Longrightarrow \Gamma^{(4)}_1. \tag{3.80}$$

3.4 Dynamics/replicas relationship

What we have seen up to now can be summarized in the following sentence: the $n \to 0$ *limit* of replica observables *is identical to* the *static limit* of the corresponding dynamical observables. We have shown this to hold in the simplest context of replica symmetry/equilibrium dynamics and considering only the u_1 interaction term. Let us now look at the case where the generalized couplings are also taken into account. This means that in the original Hamiltonian of the system we consider as interaction $\ln\cosh(\phi_i + h_i)$ and keep all the quartic terms generated after the disorder average. To simplify the algebra we only consider here the couplings u_1 and u_3, this will be anyway sufficient to illustrate how the connections with statics work, and how in the dynamical context there are also indications for a vitreous transition.

- *Generalized vertices*
  We have the correspondence

$$\frac{u_1}{4!}\sum_a (\phi^a)^4 \longleftrightarrow \frac{u_1}{6}\int \mathrm{d}t\hat{\phi}(t)\phi^3(t): \tag{3.81}$$

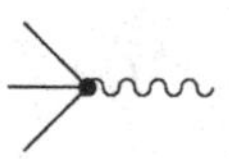

and

$$\frac{u_3}{8}\sum_a (\phi^a)^2 \sum_b (\phi^b)^2 \longleftrightarrow \frac{u_3}{2}\int \mathrm{d}t\hat{\phi}(t)\phi(t)\int \mathrm{d}t'\hat{\phi}(t')\phi(t'): \tag{3.82}$$

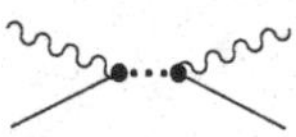

where one easily checks that the couplings have the same dimension: $[\Delta u_i] \equiv [g_i] = 6 - D$, as in the replica approach. One can generalize the dynamical perturbative expansion discussed in this chapter, taking into account all the relevant graphs arising from the presence of the new vertex.

- *Free loops*
  We have seen in the static replica approach that one could easily build free loops carrying a free summation over the replica index and hence providing an $O(n)$ contribution, e.g.

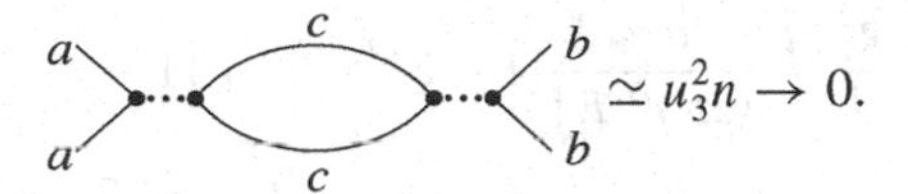

The corresponding dynamical graph is easily obtained

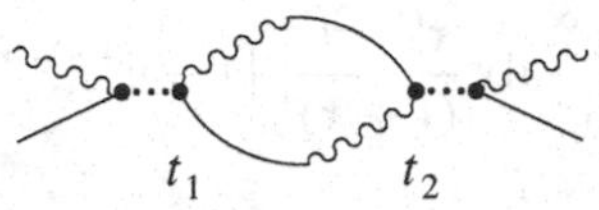

$$\simeq u_3^2 \Theta(t_1 - t_2)\, \Theta(t_2 - t_1) = 0,$$

t_1 t_2

and vanishes on account of causality. Such graphs are therefore absent in the perturbation expansion.

- *Singularities at exceptional momenta*
  In the statics, we have seen that infrared singularities arise in the generalized $\Gamma_3^{(4)}$ vertex function. Let us then look at the dynamical equivalent to check whether such singular behaviour also emerges in the dynamics. The relevant dynamical observable is then

$$\frac{1}{2}\Gamma^{(4)}_{\phi\hat{\phi}\phi\hat{\phi}}$$

$$= \quad + \quad + \cdots +$$

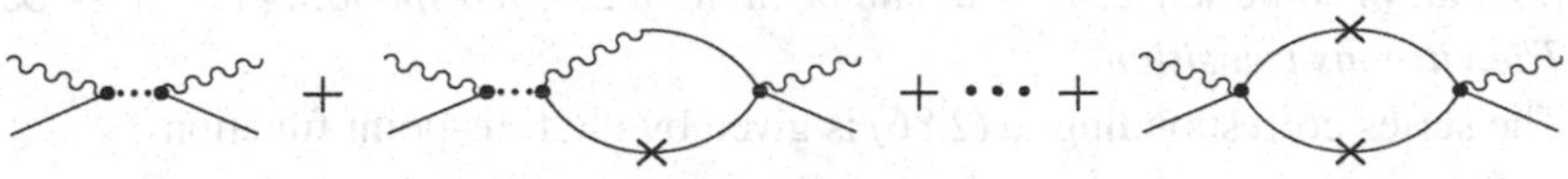

$$\frac{1}{2} \qquad \frac{1}{1} \qquad \frac{1}{4} \tag{3.83}$$

to be compared with Eq. (2.50) (but without the graph generated by the u_2 coupling).

A deeper relationship between replicas and dynamics can be extracted by considering the infrared singular contribution to $\Gamma_3^{(4)}$, what we have called D_3 in the previous chapter (see Eqs. (2.54) and (2.57)). To one loop this is given by the last graph of (2.50) and, at *exceptional momenta*, it is given by

$$D_3^0(p_i = 0) =$$

a a' b b

$$= \frac{g_1^2}{16} \int \frac{\mathrm{d}q^D}{(2\pi)^D} \frac{1}{(q^2 + r_0)^2} \frac{1}{(q^2 + r_0 - n\Delta)^2}. \tag{3.84}$$

At $r_0 = n\Delta$ (i.e. as $n \to 0$, approaching the Curie line) we have

$$D_3^0 = \frac{g_1^2}{16} \int \frac{\mathrm{d}q^D}{(2\pi)^D} \frac{1}{(q^2 + n\Delta)^2} \frac{1}{q^4} \sim \frac{g_1^2}{(n\Delta)^{(8-D)/2}}, \tag{3.85}$$

which gives a singular contribution in dimension $D < 8$.

Also, dynamically, one can see a similar singular behaviour arising when approaching the Curie line. To do that, one has to work *keeping* initially t_0 *finite*, that is using Eq. (3.70) to evaluate the last term of (3.83) at $p_i = 0$. One gets

$$\begin{aligned} D_3^0(t, t'; t_0) &= \frac{g_1^2}{4} \int \frac{\mathrm{d}q^D}{(2\pi)^D} \frac{1}{E_q^4} [1 - \mathrm{e}^{-E_q(t-t_0)}]^2 [1 - \mathrm{e}^{-E_q(t'-t)}]^2 \\ &= \frac{g_1^2}{4} \int \frac{\mathrm{d}q^D}{(2\pi)^D} \frac{1}{E_q^4} \left[\int_0^{t-t_0} \mathrm{d}\tau \, \mathrm{e}^{-E_q \tau} \right]^2 \left[\int_0^{t'-t_0} \mathrm{d}\tau' \mathrm{e}^{-E_q \tau'} \right]^2 \\ &\simeq (-t_0)^{(8-D)/2} \, y^{(D-8)/2} \int_0^\infty \mathrm{d}x \, \frac{x^{(D-2)/2}}{(1+x)^4} \left[1 - \mathrm{e}^{-y(1-x)} \right]^4, \end{aligned} \tag{3.86}$$

with $y = (-t_0)\, r_0$. Therefore for finite but large negative t_0, i.e. *when one enters the critical regime with* $r_0 \sim 1/|t_0|$ (and y finite $\sim O(1)$), the four-point function $\Gamma_3^{(4)}(t, t'; t_0)$ via its D_3 component receives contributions of order

$$|t_0|^{(8-D)/2} \tag{3.87}$$

independent of t and t'. Just like the corresponding vertex in the replica approach, (3.85) receives a contribution in

$$(n\Delta)^{(D-8)/2}\,. \tag{3.88}$$

So that, in some sense, $n \to 0$, *can be thought as corresponding to* $|t_0| \to \infty$.

- *The vitreous transition*

The series corresponding to (2.86) is given by the four-point function

$$\langle \phi(\omega)\, \phi(-\omega)\, \hat{\phi}(\omega)\, \hat{\phi}(-\omega) \rangle$$

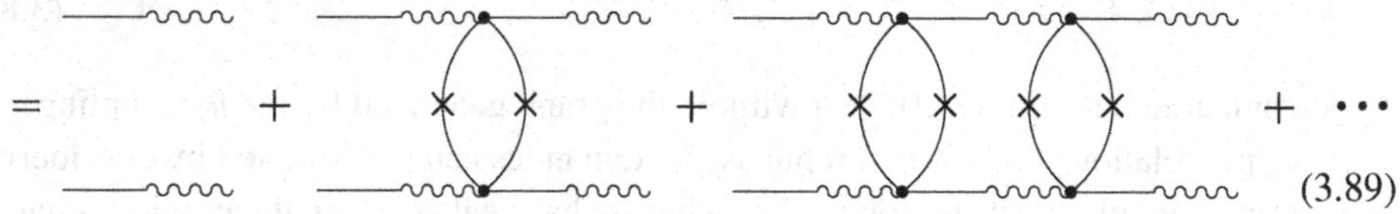

$$= \quad + \quad + \quad + \cdots \tag{3.89}$$

The static limit (where $\omega = 0$ for the external response lines) for this expression is then identical to (2.86).

3.5 Summary

In this chapter we have analyzed the dynamical behaviour of the Random Field Ising Model using the Martin–Siggia–Rose generating functional approach. This has given us a chance to introduce dynamical techniques mostly used to address Langevin dynamics for disordered systems (other applications of this method will be given in the next chapters).

We have focused on equilibrium dynamics, that is we have considered the asymptotic dynamical regime ($t_0 \to -\infty$ or, equivalently, $t \to \infty$) above the spin glass phase, where time-translation invariance can be assumed and the Fluctuation–Dissipation Theorem relating correlation and response functions does hold. What we have seen can be summarized with the following main points.

- Since the generating functional $\hat{Z}$ is by construction normalized, one does not need to introduce replicas when computing disorder-averaged dynamical observables.
- When considering only the standard quartic coupling $u_1\phi^4$, the effective dynamical action of the RFIM has one extra quadratic term, absent in the pure case, which comes from the random field average. The bare correlation function accordingly acquires a zero-frequency mode proportional to the disorder strength.
- The perturbative expansion can easily be computed, and the static limit, for which we gave the appropriate procedure, gives back the static results obtained in the previous chapter.
- A comparison between the static and the dynamical approach reveals that the role played in the statics by replica indices is in the dynamical context played by time variables. Multi-replica sums do then correspond to multi-time integrals. The dynamical counterpart of the generalized couplings appearing in the statics is given by generalized multi-time vertices.
- Infrared divergences occurring in four-point static vertex functions taken at exceptional momenta, in the limit $n \to 0$, correspond to divergences occurring in the analogous four-point dynamical functions in the limit $t_0 \to -\infty$.

References

Bouchaud J.-P., Cugliandolo L. F., Kurchan J. and Mézard M. (1998). In *Spin Glasses and Random Fields*, ed. A. P. Young, Singapore, World Scientific, p. 161.
Brézin E. and De Dominicis C. (1999). *C. R. Acad. Sci. II*, **327**, 383.
De Dominicis C. (1975). *Lett. Nuovo Cimento*, **12**, 567.
De Dominicis C. (1978). *Phys. Rev. B*, **18**, 4913.
Gardiner C. W. (1983). *Handbook of Stochastic Methods*, Berlin, Springer.
Kurchan J. (1992). *J. Phys. I France*, **2**, 1333.
Ma S. (1976). *Modern Theory of Critical Phenomena*, New York, Benjamin.
Martin P. C., Siggia E. and Rose H. (1973). *Phys. Rev. A*, **8**, 423.
Parisi G. (1988). *Statistical Field Theory*, Reading, Mass., Perseus Books, Chapter 19.
Zwanzig R. (2001). *Nonequilibrium Statistical Mechanics*, New York, Oxford University Press.

4

The $p = 2$ spherical model

In the previous chapters we have seen how the presence of disorder may be responsible for a novel behaviour at low temperature where bound states appear. In the case of the Random Field Ising Model we have been able to trace evidence of this nontrivial spin glass phase within a field theory, in both approaches from statics and dynamics. However, we did not go far enough to describe this phase and characterize its properties. In the rest of this book our aim will be precisely to address this problem, restricting ourselves to systems where it has been mostly studied in the last thirty years: spin glasses. For spin glasses, contrary to the RFIM, there is no random-site magnetic field; instead, the heterogeneity occurs in the exchange interactions between the spins that are then modelled as quenched random variables.

In this chapter we consider a first simple model of spin glass, the $p = 2$ spherical model, that is constrained to spins interacting by pairs. As usual, understanding is greatly helped if one is able to obtain an *exact* solution for a model that possesses some of the characteristic features of interest. This is the case of this spin model that, despite its coupling randomness, is exactly soluble, both for the statics and the dynamics, and does not require the replica method.

As we shall see, the $p = 2$ spherical model has not a true spin glass behaviour and is rather a disguised ferromagnet. It, however, exhibits some interesting dynamical features, such as aging and weak ergodicity breaking, that also characterize true spin glass dynamical behaviour. Since we will not address in detail the spin glass dynamics in the rest of this book, this is then a good occasion to illustrate these important concepts in a simple and understandable case.

4.1 The model

The Hamiltonian of the model is given by

$$\mathcal{H} = -\frac{1}{2}\sum_{i,j} J_{ij} S_i S_j, \tag{4.1}$$

where the spins are continuous variables obeying the global constraint

$$\sum_{i=1}^{N} S_i^2 = N. \tag{4.2}$$

The couplings J_{ij} are quenched random variables governed by a Gaussian probability distribution

$$\mathcal{P}(J_{ij}) \sim \mathrm{e}^{-N J_{ij}^2/2J^2} \tag{4.3}$$

of variance J^2/N (Kosterlitz *et al.*, 1976; Shukla and Singh, 1981).

4.2 Statics

Let us start by considering the static behaviour of this model. The partition function reads

$$Z = \int \left(\prod_i \mathrm{d}S_i\right) \exp\left\{+\frac{\beta}{2}\left[\sum_{i,j} J_{ij} S_i S_j - \ell \sum_i (S_i^2 - 1)\right]\right\}, \tag{4.4}$$

where ℓ is a Lagrange multiplier introduced to take account of the spherical constraint (4.2). It will be determined self-consistently by imposing that the constraint does hold (alternatively we could have used a delta function to directly reduce the space of the spin integration variables).

It is convenient to change basis and work with one basis that diagonalizes the set of randomly chosen J_{ij}s. Then, if we call J_λ the eigenvalues of the interaction matrix J_{ij}, we get

$$Z = \int \prod_\lambda (\mathrm{d}S_\lambda) \exp\left\{\frac{\beta}{2}\left[\sum_\lambda J_\lambda S_\lambda^2 - \ell \sum_\lambda (S_\lambda^2 - 1)\right]\right\}, \tag{4.5}$$

where S_λ indicates the projection of the spin vector $\{S_i\}$ onto the eigenvector corresponding to J_λ.

Hence, we have

$$\begin{aligned} Z &= \prod_\lambda \left(\int \mathrm{d}S_\lambda \, \exp\left\{-\frac{\beta}{2}(\ell - J_\lambda) S_\lambda^2 + \frac{\beta N \ell}{2}\right\}\right) \\ &= \exp\left\{-\frac{1}{2}\sum_\lambda \ln \beta(\ell - J_\lambda) + \frac{\beta \ell N}{2}\right\} \equiv \mathrm{e}^{-\beta f N}. \end{aligned} \tag{4.6}$$

The value of the Lagrange parameter ℓ can be fixed by considering the thermal average of constraint (4.2), that is, in the diagonal basis,

$$\sum_\lambda \langle S_\lambda^2 \rangle = N. \tag{4.7}$$

From Eq. (4.6) we then immediately get

$$1 = \frac{1}{N} \sum_\lambda \frac{T}{\ell - J_\lambda}. \tag{4.8}$$

4.3 The Wigner distribution

In order to solve Eq. (4.8) and get the value of ℓ, or to compute explicitly the free energy and other average values, we need to know how the eigenvalues J_λ are distributed. Indeed, given a generic function f, we have

$$\begin{aligned} \frac{1}{N} \sum_\lambda f(J_\lambda) &= \int d\mu \left[\frac{1}{N} \sum_\lambda \delta(\mu - J_\lambda) \right] f(\mu) \\ &\equiv \int d\mu \, \rho(\mu) f(\mu), \end{aligned} \tag{4.9}$$

where we have introduced the density of the eigenvalues $\rho(\mu)$.

This eigenvalue density can be computed in various ways, using standard methods of random matrix theory (Mehta, 1991), or even the replica method. We note that if we are interested in the average behaviour with respect to the quenched disorder, we only need the typical eigenvalue density and can disregard sub-leading contributions that vanish in the thermodynamic limit (see, however, Cavagna *et al.*, 1998 for a case where sub-leading tails give interesting physical information).

Here, we follow a simple minded expansion method. We first express the eigenvalue density in terms of the so-called 'resolvent operator'

$$\begin{aligned} \rho(\mu) &= \frac{1}{N} \sum_\lambda \delta(\mu - J_\lambda) = \frac{1}{\pi N} \operatorname{Im} \sum_\lambda \frac{1}{\mu - i0 - J_\lambda} \\ &= \frac{1}{\pi N} \operatorname{Im} \operatorname{tr} \frac{1}{z\mathbb{I} - \mathbb{J}} = \frac{1}{\pi N} \operatorname{Im} \sum_j \mathcal{G}_{jj}(z), \end{aligned} \tag{4.10}$$

where $z = \mu - i0$. Then, we expand the resolvent $\mathcal{G}$ with respect to z:

$$\mathcal{G}_{jj}(\mu) = \left(\frac{1}{z\mathbb{I} - \mathbb{J}} \right)_{jj} = \frac{1}{z} + \sum_k \frac{1}{z} J_{jk} \frac{1}{z} J_{kj} \frac{1}{z} + \cdots \tag{4.11}$$

Graphically:

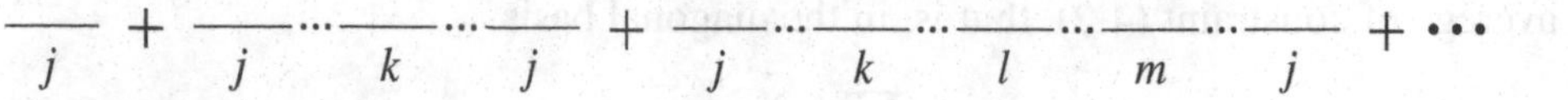

j + j ... k ... j + j ... k ... l ... m ... j + $\cdots$

(4.12)

with straight lines standing for $1/z$ factors and dotted lines for matrix elements. In this expansion, only *even* powers of J survive summations. The leading contribution of the expansion is then given by terms where a matrix element always appears *squared*. So, as a first step, one obtains

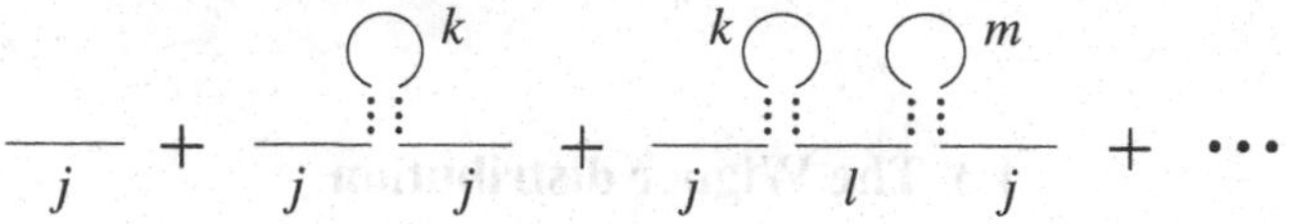

j + j (k) j + j (k) l (m) j + $\cdots$ (4.13)

or, equivalently,

$$\mathcal{G}_{jj}(z) \sim \frac{1}{z - J^2 \frac{1}{z}}, \tag{4.14}$$

where $\sum_k J_{ik}^2 \simeq (J^2/N) \sum_k = J^2$ has been used. Note that $\mathcal{G}_{jj}$ is now site independent so that we can drop the site index. A further contribution can be obtained if we decorate each internal line $1/z$:

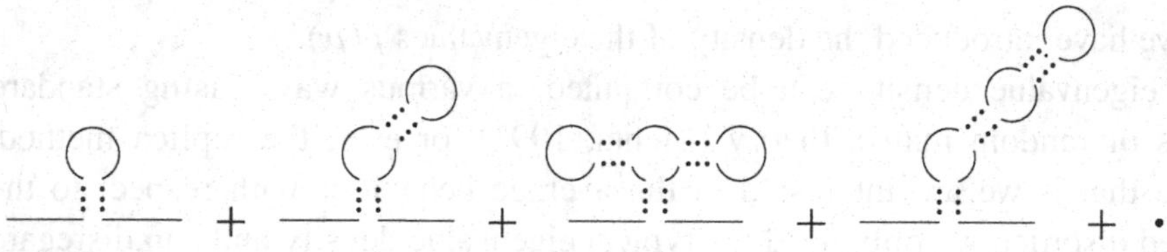

+ + + + $\cdots$,

(4.15)

thus yielding a self-consistent equation for $\mathcal{G}(z)$,

$$\mathcal{G}(z) = \frac{1}{z - J^2 \mathcal{G}(z)}. \tag{4.16}$$

†The self-consistent equation yields

$$\mathcal{G}(z) = \frac{1}{2J^2}\left[z \pm \sqrt{z^2 - 4J^2}\right]. \tag{4.17}$$

Thus we have

$$\rho(\mu) = \begin{cases} \frac{1}{2\pi J^2}\sqrt{4J^2 - \mu^2}, & \mu \le 2J, \\ 0, & \mu > 2J, \end{cases} \tag{4.18}$$

† This pedestrian 'proof' is just the manifestation of a general theorem ('t' Hooft) about planar graphs yielding the leading contribution.

which is known as the Wigner semi-circle law. Note that the disorder realization no longer appears in the expression of $\rho(\mu)$. Indeed, the eigenvalue density being a self-averaging quantity, it is sample independent at leading order. For the same reason, average values computed with (4.18) must be understood as including the disorder average.

Using now the Wigner distribution, Eq. (4.8) becomes

$$\frac{1}{T} = \int_{-2J}^{+2J} \frac{d\mu}{2\pi J^2} \sqrt{4J^2 - \mu^2} \frac{1}{\ell - \mu}. \tag{4.19}$$

4.4 A disguised ferromagnet

Equation (4.19) gives in principle the value $\ell(T)$ for any given temperature, with ℓ large if $T/J^2 \gg 1$ and ℓ decreasing with T. However, it cannot go below the critical value $\ell_c = 2J$ since otherwise the integral would become negative. This identifies the critical temperature T_c:

$$\frac{1}{T_c} = \int_{-2J}^{+2J} \frac{d\mu}{2\pi J^2} \sqrt{4J^2 - \mu^2} \frac{1}{2J - \mu}. \tag{4.20}$$

Below T_c, the weight of the borderline eigenvalue $\mu^* = 2J$ has to become *macroscopic* to allow for the constraint to be satisfied and there is a kind of Bose condensation at $\mu^* = 2J$. In other words, while at high temperatures each term in the sum (4.7) is of order one, below the critical temperature the term corresponding to the largest eigenvalue becomes extensive and accounts by itself for a certain macroscopic fraction of the whole sum. To determine quantitatively this occupation fraction it is convenient to explicitly distinguish in Eq. (4.19) the contribution at T_c:

$$\frac{1}{T} = \frac{1}{T_c} + \left(\frac{1}{T} - \frac{1}{T_c}\right) = \int_{-2J}^{+2J} d\mu\, \rho(\mu) \frac{1}{2J - \mu} + \left(\frac{1}{T} - \frac{1}{T_c}\right). \tag{4.21}$$

Here, the integral corresponds to the contribution of the whole eigenvalue spectrum without the border, while the other term corresponds to the contribution of the condensed state. Therefore, we have

$$\begin{aligned} \frac{1}{T_c} &= \frac{1}{NT} \sum_{\mu \neq \mu^*} \langle S_\mu^2 \rangle, \\ \frac{1}{T} - \frac{1}{T_c} &= \frac{1}{NT} \langle S_{\mu^*}^2 \rangle, \end{aligned} \tag{4.22}$$

or, equivalently,

$$Q_{EA} = \frac{1}{N} \langle S_{\mu^*}^2 \rangle = \left(1 - \frac{T}{T_c}\right), \tag{4.23}$$

where Q_{EA} is the so-called *self-overlap* of the condensed state. We thus have a disguised ferromagnet with macroscopic occupation of the level with $\mu = \mu^* = 2J$. The disorder is not strong enough for this model to determine a nontrivial structure of equilibrium states, as is the case for other mean field spin glasses. On the contrary, at low temperature we just find two equilibrium states (corresponding to $S_{\mu^*} = \pm\sqrt{N Q_{\mathrm{EA}}}$) related by inversion symmetry, exactly as in a ferromagnet. Note that, as in a pure Ising Model, the total magnetization computed with the global Boltzmann measure is null, $\langle S_{\mu^*} \rangle = 0$, unless degeneracy is raised by an infinitesimal magnetic field. This field should, however, be *staggered* and *sample-dependent*, since, even if the maximum eigenvalue is not sample dependent, the associated eigenvectors are. On the other hand, the self-overlap Q_{EA}, as displayed in Eq. (4.23), acquires a finite value.

4.5 Statics with a random field

Let us now add to the Hamiltonian (4.1) a *quenched random field*, i.e. we consider (Cugliandolo and Dean, 1995b)

$$\mathcal{H} = -\frac{1}{2}\sum_{i,j} J_{ij} S_i S_j - \sum_i h_i S_i, \tag{4.24}$$

with the h_i governed by a Gaussian probability law

$$\overline{h_i} = 0, \qquad \overline{h_i h_j} = \delta_{i;j}\Delta. \tag{4.25}$$

Turning to the λ-basis, we have

$$\mathcal{H} = -\frac{1}{2}\sum_\lambda J_\lambda S_\lambda^2 - h_\lambda S_\lambda, \tag{4.26}$$

and

$$\overline{h_\lambda} = 0, \qquad \overline{h_\lambda h_\mu} = \delta_{\lambda;\mu}\Delta. \tag{4.27}$$

Proceeding as in the previous section, we easily find

$$Z = \exp\left\{-\frac{1}{2}\sum_\lambda \ln \beta(\ell - J_\lambda) + \frac{1}{2T}\sum_\lambda h_\lambda \frac{1}{\ell - \lambda} h_\lambda + N\frac{\beta\ell}{2}\right\}, \tag{4.28}$$

and the new constraint condition reads

$$1 = \frac{1}{N}\sum_\lambda \left\{\frac{T}{\ell - J_\lambda} + \frac{h_\lambda^2}{(\ell - J_\lambda)^2}\right\}. \tag{4.29}$$

In terms of the eigenvalue distribution this becomes:

$$\frac{1}{T} = \int_{-2J}^{+2J} \mathrm{d}\mu\, \rho(\mu) \left\{ \frac{1}{\ell - \mu} + \frac{\Delta}{T} \frac{1}{(\ell - \mu)^2} \right\}. \tag{4.30}$$

With $\rho(\mu)$ given by the Wigner semi-circle law the Δ/T term diverges when $\ell = 2J$, thus rejecting T_c at $T_c = 0$, and suppressing the transition that generates a macroscopic occupation of the borderline eigenstate $\mu^* = 2J$.

4.6 Langevin dynamics

So far we have seen that the statics of the $p = 2$ model does not show any nontrivial phase at low temperature. One would then expect that the dynamical behaviour would be also trivial. This is only partly true (Cugliandolo and Dean, 1995a; Ciuchi and De Pasquale, 1988; Zippold, Kuhn and Horner, 2000). Indeed, we shall see that for most initial conditions the system is not able to equilibrate and will show patterns typical of off-equilibrium dynamics.

We consider, as for the RFIM, a Langevin dynamics (see Chapter 3). The equation of motion, directly written in diagonal basis, reads

$$\frac{\delta S_\lambda(t)}{\delta t} = [J_\lambda - \ell(t)]\, S_\lambda(t) + \eta_\lambda(t), \tag{4.31}$$

where η is a Gaussian noise with

$$\langle \eta_\lambda(t) \rangle = 0,$$
$$\langle \eta_\lambda(t) \eta_\mu(t') \rangle = 2T\delta(t - t')\delta_{\lambda;\mu}. \tag{4.32}$$

Now the Lagrange multiplier depends on time, since the spherical constraint has to be enforced at each time. As discussed in Chapter 3, the solution of (4.31) is given by

$$S_\lambda(t) = S_\lambda(t_0)\, \mathrm{e}^{J_\lambda t - \int_{t_0}^{t} \mathrm{d}s\, \ell(s)} + \int_{t_0}^{t} \mathrm{d}s\, \mathrm{e}^{J_\lambda(t-s) - \int_s^t \mathrm{d}s' \ell(s')} \eta_\lambda(s). \tag{4.33}$$

Unless we wish to look at the equilibrium limit and send $t_0 \to -\infty$, it is convenient to set $t_0 = 0$. It is also convenient to define

$$\gamma(t) \equiv \mathrm{e}^{2Jt - \int_0^t \mathrm{d}s\, \ell(s)}, \tag{4.34}$$

then

$$S_\lambda(t) = S_\lambda(0)\, \mathrm{e}^{-(2J - J_\lambda)t} \gamma(t) + \int_0^t \mathrm{d}s\, \mathrm{e}^{-(2J - J_\lambda)(t-s)} \eta_\lambda(s) \frac{\gamma(t)}{\gamma(s)}. \tag{4.35}$$

Imposing the constraint at (each) time t implies

$$1 = \int_{-2J}^{+2J} \mathrm{d}\mu\, \rho(\mu) \langle S_\mu^2(t) \rangle. \tag{4.36}$$

Using (4.31) and (4.32) we get

$$1 = \int_{-2J}^{+2J} \mathrm{d}\mu\, \rho(\mu) \left\{ S_\mu^2(0)\, \mathrm{e}^{-2(2J-\mu)t} \gamma^2(t) + 2T \int_0^t \mathrm{d}s\, \mathrm{e}^{-2(2J-\mu)(t-s)} \frac{\gamma^2(t)}{\gamma^2(s)} \right\}. \tag{4.37}$$

This is an equation for $D(t) = 1/\gamma^2(t)$. To solve it we need to specify more precisely the *initial conditions* $S_\mu(0)$. If $T > T_c$, we are in the paramagnetic phase, and the dynamics reduces to a simple exponential relaxation toward equilibrium. But if $T < T_c$, more interesting phenomena emerge.

4.6.1 Equilibrium initial conditions

In equilibrium we have macroscopic occupation of two 'pure' states at $\mu^* = 2J$. If we choose an initial configuration highly correlated with one of these two states, for example the positive solution of

$$S_\mu^2(0) = \frac{\delta(\mu - 2J)}{\rho(2J)}, \tag{4.38}$$

then the system is found to relax to an asymptotic value correlated with that state:

$$\begin{aligned} \langle S_{2J}(t) \rangle &\underset{t\to\infty}{\longrightarrow} +\sqrt{\left(1 - \frac{T}{T_c}\right)} S_{2J}(0) = +\sqrt{Q_{\mathrm{EA}}}\, S_{2J}(0), \\ \langle S_{\mu \neq 2J}(t) \rangle &= 0 \qquad \forall t, \end{aligned} \tag{4.39}$$

as one can see by averaging Eq. (4.35) and considering the asymptotic behaviour of Eq. (4.37).

4.6.2 Random-like initial conditions

The uniform condition $S_\mu(0) = 1$ is by all means equivalent to a *random* distribution of the initial $S_i(0)$, since the projection of the initial configuration on the equilibrium condensed one is random. The evolution is governed by Eq. (4.37) for $D(t)$,

$$D(t) = \int_{-2J}^{+2J} \mathrm{d}\mu\, \rho(\mu) \left\{ \mathrm{e}^{-2(2J-\mu)t} + 2T \int_0^t \mathrm{d}s\, \mathrm{e}^{-2(2J-\mu)(t-s)} D(s) \right\}, \tag{4.40}$$

$D(t)$ being defined as

$$D(t) \equiv \gamma^{-2}(t) = \exp\left\{ -4Jt + 2 \int_0^t \mathrm{d}s\, \ell(s) \right\}. \tag{4.41}$$

The convolution is resolved through a Laplace Transform. Let

$$\tilde{D}(z) = \int_0^\infty \mathrm{d}t \, \mathrm{e}^{-zt} D(t), \tag{4.42}$$

then Eq. (4.40) becomes

$$\tilde{D}(z) = \int_{-2J}^{+2J} \mathrm{d}\mu \, \rho(\mu) \frac{1}{z + 2(2J - \mu)} (1 + 2T \tilde{D}(z)). \tag{4.43}$$

If we define

$$\tilde{D}(z) = \frac{E(z)}{1 - 2TE(z)}, \tag{4.44}$$

then

$$\begin{aligned} E(z) &= \int_{-2J}^{+2J} \mathrm{d}\mu \, \rho(\mu) \frac{1}{z + 2(2J - \mu)} \\ &= E(0) - \frac{z}{2} \int_{-2J}^{+2J} \mathrm{d}\mu \, \rho(\mu) \frac{1}{(2J - \mu)(z + 2(2J - \mu))}, \end{aligned} \tag{4.45}$$

where we have added and subtracted $E(0) = 1/2T_{\mathrm{c}}$. To obtain the behaviour near $z \sim 0$ (i.e. $t \sim \infty$) we use

$$\int_0 \mathrm{d}u \frac{1}{\sqrt{u}} \frac{1}{u + z} \sim 1/z^{1/2},$$

so that

$$E(z) = \frac{1}{2} \left(\frac{1}{T_{\mathrm{c}}} - c z^{1/2} \right), \tag{4.46}$$

where c is a constant. Hence, finally,

$$\tilde{D}(z) = \frac{T_{\mathrm{c}}}{2} \frac{1 - c T_{\mathrm{c}} z^{1/2}}{(T_{\mathrm{c}} - T) + c z^{1/2} T T_{\mathrm{c}}} \sim \frac{T_{\mathrm{c}}}{2(T_{\mathrm{c}} - T)} \left\{ 1 - \frac{c T_{\mathrm{c}}^2}{T_{\mathrm{c}} - T} z^{1/2} + \cdots \right\}, \tag{4.47}$$

and for large t,

$$\begin{aligned} D(t) &= \frac{1}{\gamma^2(t)} \sim \frac{1}{t^{3/2}}, \\ \gamma(t) &\sim t^{3/4}. \end{aligned} \tag{4.48}$$

Knowing this explicit asymptotic behaviour, we can explore in detail various dynamical observables of interest. With slightly more effort, the solution can be computed at any given time (Cugliandolo and Dean, 1995a).

- *Magnetization*
  From Eq. (4.35), taking the noise average, we get

$$\langle S_\lambda(t)\rangle \simeq \mathrm{e}^{-2(2J-\lambda)t}\, t^{3/4}\left(1-\frac{T}{T_\mathrm{c}}\right) = \mathrm{e}^{-2(2J-\lambda)t}\, t^{3/4} Q_{\mathrm{EA}}. \tag{4.49}$$

  Thus, for $\lambda < 2J$ we find an exponential relaxation, while for the limiting eigenvalue $\lambda = 2J$ we have a $t^{3/4}$ power law, telling that the system is taking an infinite time to reach an equilibrium value (of order $\sim \sqrt{N}$).
- *Correlation*
  We have

$$\begin{aligned} C(t,t') &= \frac{1}{N}\sum_i \langle S_i(t)S_i(t')\rangle = \int_{-2J}^{+2J} \mathrm{d}\mu\,\rho(\mu)\langle S_\mu(t)S_\mu(t')\rangle \\ &= \int_{-2J}^{+2J} \mathrm{d}\mu\,\rho(\mu)\Big\{ S_\mu^2(0)\,\mathrm{e}^{-2(2J-\mu)(t+t')}\gamma(t)\gamma(t') \\ &\quad + 2T\int_0^{\min(t,t')} \mathrm{d}s\,\mathrm{e}^{-(2J-\mu)(t+t'-2s)}\frac{\gamma(t)\gamma(t')}{\gamma^2(s)}\Big\}. \end{aligned} \tag{4.50}$$

From this expression we immediately see that the correlation explicitly depends on the two times t and t', so that time translational invariance (TTI) that characterizes equilibrium dynamics does not always hold in this case. This signals, once again, that the system is not able to equilibrate even at long times. This situation is very common in disordered systems and in the following sections we try to describe it.

4.6.3 Correlation function and weak-ergodicity breaking

Let us now consider in detail the behaviour of two-time quantities. First, let us consider the correlation function and investigate how the memory of the initial condition is lost through the dynamics. To see that, one has to consider the two-point correlation function $C(t,t')$ and set $t' = 0$. We find

$$C(t,0) = \int_{-2J}^{+2J} \mathrm{d}\mu\,\rho(\mu)\,\mathrm{e}^{-(2J-\mu)t}\gamma(t). \tag{4.51}$$

This is just the average of the magnetization over the eigenvalues and it gives

$$C(t,0) \sim \frac{Q_{\mathrm{EA}}}{t^{3/4}}, \tag{4.52}$$

i.e. the memory of the initial conditions decays like a power law. One can show that a similar behaviour holds when considering the asymptotic behaviour in t for finite times t', that is when $t'/t \to 0$. In this case:

$$C(t,t') \sim f(t')\frac{1}{t^{3/4}}, \tag{4.53}$$

where f is a given function of t'. Thus the memory of any *finite* time is lost, even if very slowly, and for any given time t' there exists a large enough time t such that the correlation has decayed to zero.

Let us now consider the behaviour of $C(t, t')$ when both times are very large. In this case two different regimes may occur:

(i) *Equilibrium regime: asymptotic but close times*
This regime is attained when t and t' are both large, but relatively close one to the other, namely $t \to \infty$, $t' \to \infty$ with $(t - t')/t \ll 1$. It is convenient in this case to change variables. We set

$$t \equiv t_{\mathrm{w}} + \tau, \qquad t' \equiv t_{\mathrm{w}}, \tag{4.54}$$

where t_{w} is usually referred to as 'the waiting time'. We then let $t_{\mathrm{w}} \to \infty$, with τ/t small (which replaces the procedure of taking $t_0 \to -\infty$ for equilibrium systems). Performing this limit in Eq. (4.50) we realize that, at zero order in τ/t, time translational invariance does hold, i.e.

$$C(t, t') = C_{\mathrm{as}}(\tau). \tag{4.55}$$

The same happens with the response function

$$R(t, t') = \frac{1}{N} \sum_i \frac{\delta \langle S_i(t) \rangle}{\delta h_i(t')} \to R(\tau), \tag{4.56}$$

and the Fluctuation–Dissipation Theorem (FDT) does hold:

$$R(\tau) = -\frac{1}{T} \frac{\partial C_{\mathrm{as}}(\tau)}{\partial \tau}. \tag{4.57}$$

Besides, if we then send τ to ∞ (after t_{w}), we find

$$C_{\mathrm{as}}(\infty) = Q_{\mathrm{EA}}. \tag{4.58}$$

(ii) *Aging regime: asymptotic but widely separated times*
A very different behaviour occurs when considering well separated asymptotic times. Namely, $t \to \infty$, $t' = t_{\mathrm{w}} \to \infty$ with $(t - t_{\mathrm{w}})/t_{\mathrm{w}} = \tau/t_{\mathrm{w}} \sim O(1)$. In this case, the correlation function *depends in a nontrivial way on the two times, and not just on their difference*. The system is then said to age, since different waiting times t_{w} do correspond to different asymptotic behaviours in the time t. For example, for very separated times we find

$$C_{\mathrm{ag}}(t_{\mathrm{w}} + \tau, t_{\mathrm{w}}) \sim 2 Q_{\mathrm{EA}} \frac{\left(\frac{t_{\mathrm{w}}}{t_{\mathrm{w}}+\tau}\right)^{3/4}}{1 + \left(\frac{t_{\mathrm{w}}}{t_{\mathrm{w}}+\tau}\right)^{3/2}}. \tag{4.59}$$

Often in this regime a useful scaling variable can be found. For example, for the $p > 2$ spherical model one can show that C_{ag} only depends on the ratio t_{w}/t. In this case this

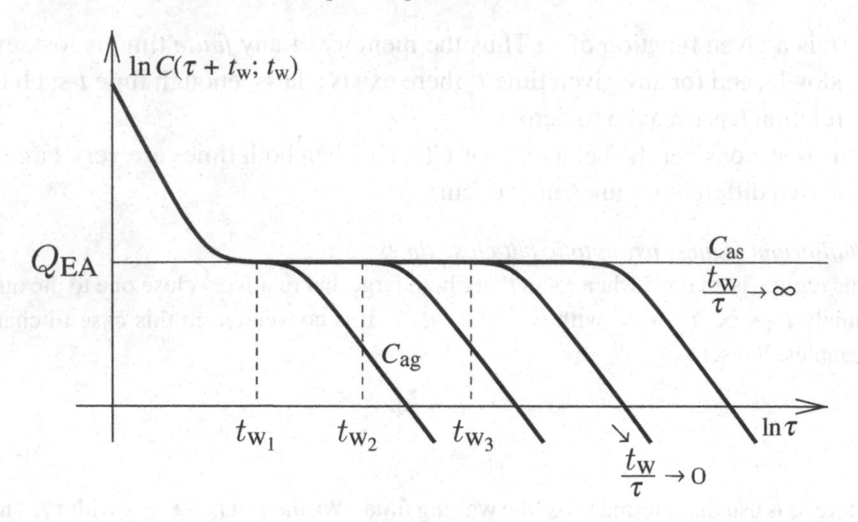

Figure 4.1 Behaviour of the correlation function at low temperature, as a function of the time difference $\tau = t - t_w$. Different curves correspond to different waiting times t_w

scaling only holds for $t_w/t \ll 1$, where (4.59) has been evaluated. Therefore, we prefer to leave explicit the dependence on t_w and t or, equivalently, t_w and τ.

For consistency the aging regime must match with the equilibrium one when τ/t_w is small, and indeed we find

$$\lim_{\tau/t_w \to 0} C_{ag} = Q_{EA} = \lim_{\tau \to \infty} C_{as}(\tau). \tag{4.60}$$

The opposite limit of this time regime is reached when $\tau \gg t_w$, i.e. when we look at a very old system on scales much larger than the waiting time. From (4.59) one finds

$$\lim_{t_w/\tau \to 0} C_{ag}\left(\frac{t_w}{\tau}\right) \sim 2Q_{EA}\left(\frac{t_w}{\tau}\right)^{3/4} \to 0. \tag{4.61}$$

Note that this limit is also recovered from Eq. (4.53) when sending t' to infinity *after* t.

In Fig. (4.1), we can see the typical dynamical pattern of the correlation function as a function of the time difference τ, at a given fixed value of t_w. For values of τ small compared to t_w the system exhibits an equilibrium-like behaviour, and seems to equilibrate at a plateau value Q_{EA}. However, when τ becomes of the order of t_w the asymptotic regime changes and the system drifts away from the plateau, ultimately decaying to zero with a power law. Note that the change of regime depends upon t_w and the age of the system: the longer we wait, the longer it will take for the correlations to decay.

So far we have seen that this system, with random initial conditions, exhibits a very peculiar behaviour. On the one hand, if we look at time scales with asymptotic but close times, it seems to equilibrate in a reduced portion of the phase space with self-overlap Q_{EA} equal to the one of the low temperature equilibrium states. However, the system never truly thermalizes since, as we have seen, on scales larger than the age of the system the correlation ultimately decays to zero. These two properties, that can be summarized as

$$\lim_{t_w \to \infty} \lim_{\tau \to \infty} C(\tau + t_w, t_w) = 0;$$
$$\lim_{\tau \to \infty} \lim_{t_w \to \infty} C(\tau + t_w, t_w) = Q_{EA}, \tag{4.62}$$

are said to characterize what has been called a *weak ergodicity breaking* scenario (Cugliandolo and Kurchan, 1993).

4.6.4 Response function and generalized FDT

Let us now turn to the behaviour of the response function. As already discussed, in the equilibrium regime it satisfies TTI and is related to the correlation function through the standard FDT. On the other hand, in the aging regime, the response explicitly depends on two times. From the dynamical equations one gets

$$R_{ag}(t_w + \tau, t_w) \sim \frac{1}{\tau^{3/2}} \left(1 + \frac{\tau}{t_w}\right)^{3/4}. \tag{4.63}$$

Again, it is useful to explore the limits of the aging scale. For $t_w \gg \tau$ we connect to the equilibrium regime. Indeed, as can be seen from (4.63), the response loses its dependence on t_w and becomes time-translational invariant. In the FDT regime, the system builds up a polarization in the eigenstate $\mu^* = 2J$. When τ increases and the system enters the aging regime, the polarization is destroyed and the correlation tends to zero. At the other end of the regime scale, $t_w/\tau \to 0$, we have

$$R_{ag} \sim \frac{1}{\tau^{3/4}} \frac{1}{t_w^{3/4}}. \tag{4.64}$$

In the aging regime the standard FDT does not hold. One may wonder whether some more general relation exists between correlation and the response function, that also describes the off-equilibrium asymptotic aging behaviour. Actually, it seems that for a large class of out-of-equilibrium systems such a relation exists.

One can define a function $X(t, t_w)$ that characterizes the violations of FDT:

$$R(t, t_w) = \frac{X(t, t_w)}{T} \frac{\partial C(t; t_w)}{\partial t_w}. \tag{4.65}$$

In the asymptotic aging regime ($t, t', \tau \gg 1$) it has been conjectured (Cugliandolo and Kurchan, 1993) and verified in several instances (numerically or otherwise) that

$$X(t, t_w) \to X(C(t, t_w)). \tag{4.66}$$

The precise behaviour of the function $X(C)$ depends on the class of systems one is considering. Examples of different behaviours can be found in disordered mean field models where $X(C)$ can be a two-step function or exhibit a continuous dependence on the correlation (Cugliandolo and Kurchan, 1994).

In the simple mean field system considered here, by comparing Eqs. (4.59) and (4.63), one finds

$$X_{\text{ag}}(t, t_w) \sim \frac{1}{t^{1/2}} \to 0. \tag{4.67}$$

Therefore, the function $X(C)$ exhibits a two-step behaviour, being equal to one in the equilibrium regime and becoming zero in the aging one:

$$\begin{aligned} X_{\text{eq}} &= 1, \\ X_{\text{ag}} &= 0. \end{aligned} \tag{4.68}$$

4.7 Connection with domain coarsening

Consider a ferromagnet with N components. In the low T phase, we have

$$\mathcal{H} = \left(\nabla \vec{S}\right)^2 - \frac{1}{2}(T_c - T)\vec{S}^2 + \frac{g}{8N}\left(\vec{S}^2\right)^2. \tag{4.69}$$

If we consider the time evolution of such a system in the low temperature phase, when started in a random configuration, we observe a phenomenon usually called *coarsening*, where domains of positive and negative magnetization increase their size in time and the system is never able to truly thermalize (at least without appropriate boundary conditions that explicitly break the symmetry). On the other hand, if the initial configuration has a macroscopic correlation with one of the two pure states (positive vs. negative magnetization), the system easily thermalizes in the corresponding state (Bray, 1994). This seems very much similar to what we have analyzed for the $p = 2$ spherical model. Indeed, we convinced ourselves that, despite the disorder, this system behaves statically as a disguised ferromagnet, so we would expect a similar behaviour also for what concerns the dynamics.

The connection can be made more quantitative if we consider the Langevin dynamics associated with Eq. (4.69):

$$\frac{\partial}{\partial t} S_a(x, t) = \Delta S_a + S_a\left((T_c - T) - \frac{g}{2N}\vec{S}^2\right) + \eta_a, \tag{4.70}$$

with $(a = 1, 2, \ldots, N)$. For large N, one can neglect fluctuations of $\vec{S}^2/N$, hence Eq. (4.70) can be rewritten as, after Fourier Transforming,

$$\frac{\partial}{\partial t} S_a(k,t) = -k^2 S_a(k,t) + (T_c - T) S_a(k,t) + \eta_a(k,t). \qquad (4.71)$$

Thus, we can recover formally Eq. (4.31) with the correspondence

$$\begin{aligned} J_\lambda \equiv \mu &\to -k^2, \\ 2J = \mu^* &\to 0, \\ \mathrm{d}\mu\sqrt{4-\mu^2} &\to k^{D-1}\mathrm{d}k, \\ T_c - T &\to \ell. \end{aligned}$$

The correspondence is complete in $D = 3$. Therefore, we expect coarsening to exhibit the same patterns of FDT violation as the ones described in the previous sections. Surprisingly, the FDT has not been analyzed for coarsening systems until recently (Barrat, 1998; Berthier, Barrat and Kierchan, 1999), and it has been shown that it actually displays the behaviour predicted by analytical arguments (Cugliandolo *et al.*, 1997), with X asymptotically zero in the aging regime.

4.8 Dynamics with a random field

So far, we have seen that the $p = 2$ spherical model is not able to equilibrate when the system is started below the critical temperature with a random initial condition. Indeed, the correlation function depends upon the two time arguments, the memory of initial conditions decays very slowly and aging appears in a well defined asymptotic regime. One would expect, on the basis of linear response theory, that a similar scenario would hold when a small magnetic field is applied to the system. As we shall now see, this is not the case.

Let us consider the $p = 2$ spherical model with a random site magnetic field. To work in a nontrivial region (i.e. nonparamagnetic) we need to take $T = T_c = 0$. The reason, as we have seen in Section 4.5, is that the static constraint equation reads in this case

$$1 = \int_{-2J}^{+2J} \mathrm{d}\mu\, \rho(\mu) \left\{ \frac{T}{\ell - \mu} + \frac{\Delta}{(\ell - \mu)^2} \right\}, \qquad (4.72)$$

and the μ integration over the Δ term is divergent for a borderline value of $\ell^* = 2J$.

If we then turn to the Langevin dynamics, and set $T = 0$, we have, instead of (4.40) and (4.41),

$$S_\mu(t) = S_\mu(0)\, \mathrm{e}^{-(2J-\mu)t} \gamma_0(t) + \int_0^t \mathrm{d}s\, \mathrm{e}^{-(2J-\mu)(t-s)} \frac{\gamma_0(t)}{\gamma_0(s)} (\eta_\mu(s) + h_\mu). \quad (4.73)$$

For the constraint (for random-like initial conditions), instead of Eq. (4.37) we have

$$1 = \gamma_0^2(t) \int_{-2J}^{+2J} d\mu\, \rho(\mu) \left\{ e^{-2(2J-\mu)t} + \Delta \left(\int_0^t ds\, e^{-(2J-\mu)(t-s)} \frac{1}{\gamma_0(s)} \right)^2 \right\}, \tag{4.74}$$

a difficult nonlinear equation. Cugliandolo and Dean (1995b) have obtained a qualitative behaviour by trying the ansatz ($\alpha > 0$)

$$\gamma_0(t) = c\, e^{-\alpha t}. \tag{4.75}$$

With this ansatz the Δ term dominates in (4.74) (for long times) over the term containing the memory of initial conditions, and we get

$$c^{-2}\, e^{2\alpha t} = \Delta \int_{-2J}^{+2J} d\mu\, \rho(\mu) \left[\frac{e^{\alpha t} - e^{-(2J-\mu)t}}{2J + \alpha - \mu} \right]^2, \tag{4.76}$$

that is

$$\begin{aligned} 1 &= \Delta c^2 \int_{-2J}^{+2J} d\mu\, \rho(\mu) \left[\frac{1 - e^{-(2J+\alpha-\mu)t}}{2J + \alpha - \mu} \right]^2 \\ &\sim \Delta c^2 \int_{-2J}^{+2J} d\mu\, \rho(\mu) \frac{1}{(2J + \alpha - \mu)^2}. \end{aligned} \tag{4.77}$$

This equation is verified for

$$\begin{aligned} \alpha &= \frac{2 + \Delta}{\sqrt{1 + \Delta}} - 2, \\ \gamma_0 &= c\, e^{-\alpha t} \equiv c\, e^{-t/\tau_0}, \end{aligned} \tag{4.78}$$

where $\tau_0 = 1/\alpha$ is a characteristic time scale. The behaviour of γ_0 (Eq. (4.78)) is qualitatively different from the one of γ (Eq. (4.48)) where the magnetic field is absent: there, we found an increasing power law behaviour, here we find an exponential decrease! As a consequence we now have a *finite* time scale, τ_0, such that for times larger than τ_0 the memory of initial conditions, a crucial ingredient for obtaining an aging behaviour, is lost. Note, however, that for $\Delta \ll 1$ this time scale becomes very large, i.e. $\tau_0 \sim 4/\Delta^2$. This means that when the applied magnetic field is small there may be a large, even if finite, regime where off-equilibrium phenomena can be observed. This can be seen also in the behaviour of the correlation function.

Instead of (4.50) we now get

$$C(t,t') = \gamma_0(t)\gamma_0(t') \int_{-2J}^{+2J} \mathrm{d}\mu\, \rho(\mu) \mathrm{e}^{-(2J-\mu)(t+t')}$$
$$+ \Delta \left[\int_0^{t'} \mathrm{d}s\, \mathrm{e}^{-(2J-\mu)(t-s)} \frac{1}{\gamma_0(s)} \right] \left[\int_0^{t'} \mathrm{d}s'\, \mathrm{e}^{-(2J-\mu)(t-s')} \frac{1}{\gamma_0(s')} \right]. \tag{4.79}$$

We have two regimes:

(i) *Aging regime* For $t, t' \ll \tau_0$ the memory of initial conditions dominates for large enough times.

(ii) *Interrupted aging* On the other hand when $t, t' \gg \tau_0$ the Δ term dominates, initial conditions are forgotten, we get $C\left(t, t'\right) \sim 1$ and we are in the equilibrium regime (since for $T = 0$ we have $Q_{\mathrm{EA}} = 1$).

The question that remains open at this point is why in this simple model the effect of a small magnetic field can be so destructive as to change qualitatively the dynamical scenario. The reason lies in the deep connections that exist for many disordered systems between the dynamical behaviour at low temperatures and the topological structure of the phase space. In general we can say that the complexity of this structure reflects itself in the complexity of the dynamical behaviour of the system. In the present case it can be shown (Cugliandolo and Dean, 1995b) that when a magnetic field is added we switch from a very rich phase space, with many saddle points and flat regions, to a trivial one. Thus the topological properties may be very fragile to external perturbations, and we give a few other interesting examples in Chapter 7 of this book.

4.9 What do we get from the replica way?

The $p = 2$ spherical model is, as we have seen, exactly solvable with standard analytical tools. This will not be the case for many other spin glass systems where one needs to extensively use the replica method. It is then interesting in this particular case, where a direct exact approach exists, to recover the static results using replicas (Kosterlitz *et al.*, 1976; Crisanti and Sommers, 1992). We start from the Hamiltonian

$$\mathcal{H} = -\frac{1}{2} \sum_{i,j} J_{ij}\, S_i\, S_j + \frac{\ell}{2} \sum_i \left(S_i^2 - 1\right) + \sum_i h_i\, S_i. \tag{4.80}$$

Within the replica method, we recall, one needs to compute the partition function of a replicated system of n copies of the original one, averaged over the quenched disorder. In this case, we must take the Gaussian averages over the random J_{ij}, and h_i. We get

$$\overline{Z^n} = \int \prod_{i,a} (\mathrm{d}s_i^a) \exp\left\{ \frac{(\beta J)^2}{4N} \sum_{i,j} \left(\sum_a S_i^a S_j^a \right)^2 + \frac{\beta^2 \Delta}{2} \sum_i \left(\sum_a S_i^a \right)^2 - \frac{\ell}{2} \sum_{i,a} (S_i^{a2} - 1) \right\}, \tag{4.81}$$

where, as usual, the overbar stands for an average over the disorder. To obtain a single site effective action, we then use the Hubbard–Stratonovich transformation

$$\frac{1}{\sqrt{2\pi a}} \int \mathrm{d}x\, \mathrm{e}^{-\frac{x^2}{2a} + xs} = \mathrm{e}^{+\frac{1}{2} a s^2}, \tag{4.82}$$

with $s = \sum_i S_i^a S_i^b$, $x = (\beta J)^2 Q_{ab}$ and $a = (\beta J)^2 / N$. We then rewrite $\overline{Z^n}$ as

$$\begin{aligned} \overline{Z^n} &= \int \prod_{i,a} (\mathrm{d}s_i^a) \prod_{(a,b)} \left(\frac{\mathrm{d}Q^{ab}}{\sqrt{2\pi}} \right) \exp\left\{ -N \frac{(\beta J)^2}{4} \sum_{a,b} (Q^{ab})^2 + \frac{(\beta J)^2}{2} \sum_{\substack{a,b \\ i}} Q^{ab} S_i^a S_i^b \right. \\ &\qquad \left. + \frac{(\beta^2 \Delta)}{2} \sum_i \left(\sum_a S_i^a \right)^2 - \frac{\beta \ell}{2} \sum_{i,a} \left((S_i^a)^2 - 1 \right) \right\} \\ &= \int \prod_{(a,b)} \left(\frac{\mathrm{d}Q^{ab}}{\sqrt{2\pi}} \right) \exp\left\{ N \left[-\frac{1}{4} (\beta J)^2 \sum_{a,b} (Q^{ab})^2 \right. \right. \\ &\qquad \left. \left. + \ln \int \prod_a \mathrm{d}s^a \exp -\frac{1}{2} \sum_{a,b} S^a M^{ab} S^b \right] \right\}, \end{aligned} \tag{4.83}$$

where subextensive constants are omitted, and

$$M^{ab} = \beta \ell\, \delta_{a;b} - \beta^2 \Delta - (\beta J)^2\, Q^{ab}. \tag{4.84}$$

Note that, according to (4.82), the variables Q_{ab} are homogeneous to spin overlaps, for this reason Q is usually called *the overlap matrix*. To go further we have to make an ansatz for the matrix Q_{ab}. Let us start with the simplest possible one, the so-called *replica symmetric* ansatz that preserves the symmetry with respect to replica permutations of the replicated Hamiltonian. We then assume

$$Q^{ab} = (1 - Q)\, \delta_{a;b} + Q. \tag{4.85}$$

Of course our choice has to be validated a posteriori by a stability study. Then Eq. (4.84) becomes

$$M^{ab} = (\beta \ell - (\beta J)^2 (1 - Q)) \delta_{a;b} - (\beta^2 \Delta + (\beta J)^2 Q). \tag{4.86}$$

It is convenient to write M in a projector form where its eigenvalues appear explicitly, as in Section 2.3 we get

$$M^{ab} = A\left(\delta_{a;b} - \frac{1}{n}\right) + \frac{1}{n}(A - nB), \tag{4.87}$$

where

$$\begin{aligned} A &= \beta\ell - (\beta J)^2 (1 - Q), \\ B &= \beta^2 \Delta + (\beta J)^2 Q. \end{aligned} \tag{4.88}$$

In Eq. (4.87) the two eigenvalues A and $A - nB$ occur with their multiplicity $(n-1)$ and 1 respectively. We now rewrite the exponent of (4.83) as

$$\begin{aligned} -\beta f N n = N \Bigg\{ & -\frac{1}{4}(\beta J)^2 \left((n-1)(1-Q)^2 + (1 - Q + nQ)^2\right) \\ & -\frac{1}{2}((n-1)\ln A + \ln(A - nB)) + \frac{\beta \ell n}{2} \Bigg\}. \end{aligned} \tag{4.89}$$

At this point we note that, in the thermodynamic limit, the integral over the Q variables in (4.83) can be performed via the saddle point approximation: it is then given by the integrand evaluated at the point that extremizes the effective action (4.89). Then, we finally get

$$-\beta f = -\frac{(\beta J)^2}{4}(1 - Q^2) - \frac{1}{2}\ln A + \frac{1}{2}\frac{B}{A} + \frac{\beta\ell}{2} + O(n), \tag{4.90}$$

where f is to be stationary with respect to Q, i.e. taken at the saddle point, and with respect to ℓ to implement the spherical constraint. Note that, at the saddle point, from Eq. (4.83) we have

$$Q_{ab} = \frac{1}{N}\sum_i \overline{\langle S_i^a S_i^b \rangle}, \tag{4.91}$$

with the average taken over the replicated Boltzmann measure.

The stationarity equations read

$$\frac{1}{A} + \frac{B}{A^2} = 1, \tag{4.92}$$

$$\frac{1}{A} + Q = 1. \tag{4.93}$$

For $\Delta = 0$, then we find the following equation for Q:

$$Q = \frac{(\beta J)^2 Q}{A^2}, \tag{4.94}$$

which has two solutions:

(i) *Paramagnetic solution*

$$\begin{cases} Q = 0, \\ \ell = 1 + (\beta J)^2 . \end{cases} \tag{4.95}$$

(ii) *Disguised ferromagnet solution*
If $Q \neq 0$ Eq. (4.94) becomes $A = (\beta J)$, and from (4.93)

$$(\beta J)^2 Q = (\beta J)^2 \left[1 - \frac{1}{\beta J} \right] \tag{4.96}$$

That is, summarizing,

$$\begin{cases} Q = 1 - \dfrac{T}{T_c}, \\ \beta \ell = 2\beta J, \end{cases} \tag{4.97}$$

with $T_c = J$. For $T > T_c$ we have only the paramagnetic solution. For $T \leq T_c$ we also have the disguised ferromagnet (with condensation on the borderline eigenvalue). This last solution is the thermodynamically relevant one despite the fact that, or rather because, it has the larger free energy. This is a consequence of the limit $n \to 0$ performed within the replica method that implies an inverted thermodynamical rule. We shall return to that later.

4.10 Comments and summary

In this chapter we have analyzed the simplest model of a system with quenched disordered interactions. The Hamiltonian of the $p = 2$ spherical model is analogous to the one of the Sherrington–Kirkpatrick model, which is considered as the archetype of spin glass behaviour. However, in the spherical case the spins are not discrete Ising variables, but are continuous and obeying a spherical constraint. This difference in the accessible configuration space entails tremendous differences when looked at from the thermodynamical viewpoint. The freedom allowed for the continuous spins is enough to compensate for the frustration effects induced by the quenched disorder, and the low temperature phase is nothing but a ferromagnet in disguise. On the other hand, as we shall see in detail in the next chapters, when one considers discrete spins or more complex p-body interactions, the properties at low temperature are much richer, with the occurrence of many equilibrium states (and not just two) in the so-called *spin glass phase*.

Despite its trivial thermodynamics, this model exhibits a nontrivial dynamical behaviour when random initial conditions are considered. Many of the features we have analyzed in this chapter are indeed typical of the off-equilibrium dynamics occurring in real spin glass systems. The general framework, where two asymptotic dynamical regimes are present, a quasi-equilibrium and an aging one, is valid for most of the glassy systems known so far. However, the aging behaviour that we find is, in some respects, a *soft* one and, in a dynamical perspective, a simpler one than that of a true spin glass. The FDT violation factor X, for example, is equal to zero in the aging regime, indicating that there is a loss of memory in the response, while spin glasses do have long-term memory, and indeed the FDT ratio acquires there a finite value. In other words, the response function decays with a power law but with an exponent that is larger than the one found in spin glasses. As a consequence, the memory is too *weak* and aging effects are washed away quickly. This can be seen by looking, for example, at the decay of the magnetization when a magnetic field is applied in the interval $[0, t_w]$ (thermoremanent magnetization):

$$M_{t_w}(t) = \int_0^{t_w} \mathrm{d}s\, R(t, s). \tag{4.98}$$

From the behaviour of the response function we get a power law decay in the two cases of finite t_w and asymptotic t_w (even if with different exponents). In this sense the memory of the system of its history, when integrated over time, is always weak and is ultimately lost. In true spin glasses, on the other hand, $M_{t_w}(t)$ still decays to zero for finite t_w but approaches a finite value for large t_w, indicating that these systems possess long-term memory, even if short-time details are forgotten (what is called *weak long-term memory*, see Cugliandolo and Kurchan, 1993). Note that, if we use the modified FDT in Eq. (4.98), having long-term memory or not is equivalent to saying that X_{ag} decays asymptotically to zero or has a finite value.

We have seen that a behaviour like that of the $p = 2$ spherical model exists in coarsening systems. A simple physical interpretation for it can be given. At long times the response of a coarsening system is dominated by the bulk response of the domains that form during the coarsening process. The response of a single spin at time t to a field applied at time t' is finite only if the spin is *not* swept by a domain wall between t and t'. The excess response is due to the domain walls, their density, however, decreases in time and this contribution therefore vanishes at long times. More quantitatively, one can look at the behaviour of the so-called *staggered* magnetization $M(t, t_w)$: i.e. one applies a random site magnetic field at time t_w and monitors how the average magnetization in the direction of the field evolves in time. The magnetization is computed by integration of the response function, and, if we assume a generalized FDT, we can express it in terms of X:

$$M(t, t_w) = \frac{1}{T} \int_{C(t,t_w)}^{C(t,t)} \mathrm{d}C\, X(C). \tag{4.99}$$

At short time differences (large C) the magnetization increases linearly with time (equilibrium regime), however, when the system enters the aging regime (larger times, i.e. smaller values of C) $X = 0$ and one observes a flattening of the staggered magnetization. Physically, this means that the system is old enough for the bulk contribution to be dominant, but this last one being time independent, the magnetization becomes stationary.

Summarizing, we may then say:

- Not all disordered systems exhibit a nontrivial thermodynamic behaviour; sometimes the spins have enough freedom to overcome frustration effects (actually the question is still open whether dimensionality can have a role in this aspect).
- Systems with a trivial thermodynamics may exhibit off-equilibrium dynamical behaviour.
- The framework of the generalized Fluctuation–Dissipation Theorem allows us to classify models according to the degree of violation of FDT. Simpler coarsening systems correspond to $X_{ag} = 0$, i.e. aging without long-term memory, while spin glasses correspond to $X_{ag} > 0$.

References

Barrat A. (1998). *Phys. Rev. E*, **57**, 3629.
Berthier L., Barrat J.-L. and Kurchan J. (1999). *Eur. Phys. J. B*, **11**, 635.
Bray A. (1994). *Adv. Phys.*, **43**, 357.
Cavagna A., Giardina I. and Parisi G. (1998). *Phys. Rev. B*, **57**, 11251.
Ciuchi S. and De Pasquale F. (1988). *Nucl. Phys. B*, **300**, 31.
Crisanti A. and Sommers H. J. (1992). *Z. Phys. B*, **87**, 341.
Cugliandolo L. F. and Dean D. (1995a). *J. Phys. A*, **28**, 4213.
Cugliandolo L. F. and Dean D. (1995b). *J. Phys. A*, **28**, L453.
Cugliandolo L. F. and Kurchan J. (1993). *Phys. Rev. Lett.*, **71**, 173.
Cugliandolo L. F. and Kurchan J. (1994). *J. Phys. A*, **27**, 5749.
Cugliandolo L. F., Kurchan J. and Peliti L. (1997). *Phys. Rev. E*, **55**, 3898.
Kosterlitz J., Thouless D. J. and Jones R. (1976). *Phys. Rev. Lett.*, **36**, 1217.
Mehta M. L. (1991). *Random Matrices*. Revised and enlarged 2nd edition, San Diego, Academic Press.
Shukla P. and Singh S. (1981). *Phys. Rev. B*, **23**, 4661.
Zippold W., Kuhn R. and Horner H. (2000). *Eur. Phys. J. B*, **13**, 531.

5

Mean field spin glasses: one-step RSB

Let us now consider mean field models of spin glasses where, contrary to the simple case studied in the previous chapter, the quenched disorder dramatically affects the thermodynamics. For these models, as we shall see, the replica method not only represents a necessary mathematical tool to perform the computations, but also offers an appropriate description of the nontrivial low temperature physics.

When looking at the thermodynamics of the $p = 2$ spin model using replicas we ended up with a Lagrangian (Eq. (4.83)) expressed in terms of an overlap matrix Q_{ab}. To perform the calculation we had to make an ansatz on the structure of this replica matrix, and we chose the simplest possible one, the *Replica Symmetric* (RS) ansatz. In that case, this ansatz proved to be appropriate, however, this is in general not true: whenever real spin glass behaviour is encountered the RS solution is not the appropriate one, being unstable and leading to unphysical predictions (such as, for example, negative entropy). Alternative ansatzes for the Q_{ab} matrix necessarily involve a breaking of the replica symmetry, and are usually referred to as *Replica Symmetry Broken* (RSB) solutions.

The correct way to break the symmetry was the main focus of most theoretical research on spin glasses in the late seventies. As it turned out later on, the degree of symmetry breaking necessary to find a stable solution actually depends on the model considered. There are models where one needs to break the symmetry just once, the so-called 'one-step RSB' ansatz. There are others where a continuous infinite hierarchy of breakings is required, with the so-called 'full RSB' ansatz.

Historically, the one-step RSB was introduced as a first attempt to solve the Sherrington–Kirkpatrick (SK) model (Sherrington and Kirkpatrick, 1975). Indeed, Sherrington and Kirkpatrick in their paper used a replica symmetric ansatz which turned out to be unstable and yielding besides a negative entropy. Many tries were then made to improve the result, until, as we shall see in the next chapter, Parisi realized that a one-step RSB was rendering the entropy less negative. By a bold generalization (to an infinite number of steps) he displayed a solution with non-negative entropy that was later to be proven stable and indeed exact.

In this chapter we address another class of models for which one-step ansatzes represent the correct solution, in contrast to what is said above. The simplest of such models is the Random Energy Model (REM) introduced by Derrida (Section 5.1) below. A more complex model under scrutiny (Section 5.2) will be the p-spin model with $p > 2$ (p-plets couplings instead of the pair-couplings of $p = 2$). The spins can be discrete as for Ising spins or continuous variables with a sphericity constraint as in the previous chapter (Crisanti and Sommers, 1992). For $p > 2$ the Ising and the spherical cases display anyway the same phenomenology. The case $p = 2$ with Ising spins, i.e. the Sherrington–Kirkpatrick Model, is treated in the following chapters.

Apart from being a good introductory exercise for the more complicated SK case, there are two good reasons why models solved by a one-step RSB are of interest. From a mathematical viewpoint, it will turn out that there is a deep relationship between the one-step RSB solution and the statistics of extremes (Section 5.1.1). Besides, as we discuss in Section 5.2.6, these models are also of interest physically speaking since it has recently been realized that they provide a mean field description for structural glasses (while, on the other hand, they are not adequate to describe real spin glasses). In the following we also briefly discuss these two issues, referring the reader to some relevant further reading for a deeper and more detailed analysis.

5.1 The Random Energy Model (REM)

The REM is the simplest possible case where one-step RSB is exhibited, and it can also be studied without using replicas. It was introduced by Derrida (Derrida, 1981; Cabibbo, 1979, unpublished) and is defined in the following way.

There are $M \equiv 2^N$ configurations whose energies $\{E_i\}$ are independent random variables, identically distributed according to the probability distribution

$$P(E_i) = A\, \mathrm{e}^{-\frac{E_i^2}{N}}, \tag{5.1}$$

where $i = 1, \ldots, M$ and A is the normalization factor. For this model the quenched disorder does not appear through random exchange interactions but as random energy levels. In this case then a 'sample', that is a particular realization of the quenched disorder, is represented by a set $\{E_i\}$ of random variables.

For a given sample the partition function is:

$$Z = \sum_{i=1}^{2^N} \mathrm{e}^{-\beta E_i}. \tag{5.2}$$

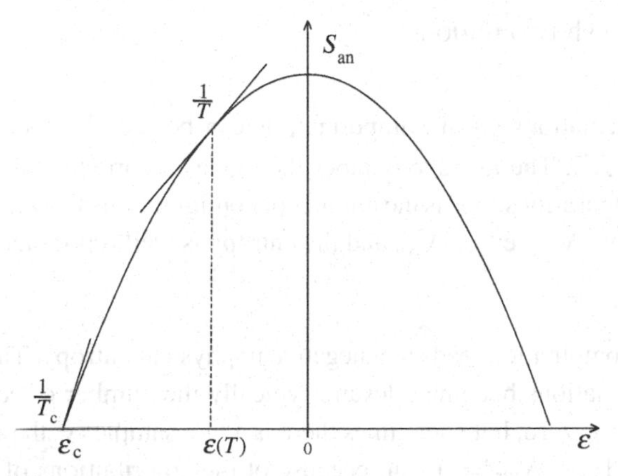

Figure 5.1 Annealed entropy as a function of the energy density. The straight line corresponds to a derivative $1/T$ and is used to evaluate the equilibrium energy density at temperature T, see text

The REM can be solved using *a microcanonical approach*, without computing the partition function, but we will come back to (5.2) later on. In a microcanonical context one needs to compute the entropy $S(E)$ related to the number $\mathcal{N}(E)$ of levels with energy E. The other equilibrium properties can then be obtained using standard thermodynamic relations. The system being disordered, we must consider averages over the disorder distribution (indicated, as usual, with overbars). As previously discussed in the Introduction, averages should be performed on self-averaging observables (e.g. extensive ones), since for these quantities average values coincide with typical ones. On the contrary, for nonself-averaging quantities the contribution of rare samples may lead to estimates that do not reproduce the typical behaviour of the ensemble. In the present case, this means that we should compute the typical entropy performing the *quenched* average $S(E) = \overline{\ln \mathcal{N}(E)}$, rather than the *annealed* one $S_{\mathrm{an}}(E) = \ln \overline{\mathcal{N}(E)}$. To illustrate the difference between these two averages, let us consider first the annealed one, which is simpler to perform. We have

$$\overline{\mathcal{N}(E)} = \sum_{i=1}^{2^N} \overline{\delta(E_i - E)} = 2^N P(E) = 2^N A\, \mathrm{e}^{-E^2/N}. \tag{5.3}$$

Introducing the intensive variable $\varepsilon = E/N$ we get

$$\overline{\mathcal{N}(E)} = A\, \mathrm{e}^{N[\ln 2 - \varepsilon^2]}. \tag{5.4}$$

There is then a finite entropy related to this average number, for energy densities $\varepsilon \in [\varepsilon_c; -\varepsilon_c]$, with $\varepsilon_c = -\sqrt{\ln 2}$. This can be seen more explicitly in Fig. 5.1, where we plot $\ln \overline{\mathcal{N}(E)}/N$ as a function of ε.

We can distinguish two regions:

(i) $\varepsilon^2 < \varepsilon_c^2$
In this case fluctuations are not important, as can be seen by looking at the second moment $\overline{\mathcal{N}^2} \sim \overline{\mathcal{N}}^2$. The average number then gives a correct estimate of the *typical* number of configurations, that is the number of configurations in a highly probable sample. We thus have $\overline{\mathcal{N}} \sim e^N \sim \mathcal{N}_{typ}$ and the entropy is well reproduced by the annealed computation.

(ii) $\varepsilon^2 > \varepsilon_c^2$
The annealed computation leads to a negative unphysical entropy. The reason is that in this region fluctuations become relevant: typically the number of configurations with energy density ε is zero, but sometimes there is a rare sample (with probability $\sim e^{-N}$) where $\mathcal{N} \neq 0$. Thus, $\mathcal{N}_{typ} = 0$, but, because of the contributions of the rare samples, $\overline{\mathcal{N}} \sim e^{-N}$.

Let us now look at the thermodynamics. The entropy is given by $S = \overline{\ln \mathcal{N}(E)} = \ln \mathcal{N}_{typ}$. For $\varepsilon^2 < \varepsilon_c^2$ the annealed computation is correct, giving for the intensive entropy $s = S/N = \ln 2 - \varepsilon^2$. At equilibrium we must have $1/T = \mathrm{d}s/\mathrm{d}\varepsilon$. That is, at a given temperature T, the equilibrium energy density $\varepsilon(T)$ is given by the point in the curve s versus ε where the derivative is equal to $1/T$ (i.e. $\mathrm{d}s/\mathrm{d}\varepsilon = -2\varepsilon = 1/T$, see Fig. 5.1). From Fig. 5.1 we can see that there is a temperature T_c where $\varepsilon(T_c) = -1/(2T_c) = \varepsilon_c$. Below this temperature one cannot continue looking at the derivative along the curve because it would enter in the region $\varepsilon < \varepsilon_c$ where s, as computed in the annealed way, is negative. Since, typically, there are no levels with $\varepsilon < \varepsilon_c$, the best that the system can do to comply with thermodynamical constraints is to choose the lowest possible energy. Therefore, at T_c there is a phase transition, and the system freezes at the lower band edge in levels with energy density ε_c and entropy $s = 0$. These levels remain the dominant ones in the whole low temperature phase. If we now consider the free energy density $F/N = \varepsilon(T) - Ts(T)$, we then find

$$F/N = \begin{cases} -T\ln 2 - \dfrac{1}{4T}; & T > T_c, \\ -\sqrt{\ln 2}; & T < T_c. \end{cases} \tag{5.5}$$

Another way to get the same results is to look directly at the partition function. Again, Z being nonself-averaging, we should in principle compute $\overline{\ln Z}$, which is what we shall do with the replica method in one of the next sections. However, from the previous analysis we understood that the annealed average fails at low temperature because it badly estimates the number of levels with $\varepsilon < \varepsilon_c$. Keeping this in mind, we can heuristically compute Z by attributing the correct number density to these levels, i.e. $\mathcal{N}_{typ}(\varepsilon) = 0$ for $\varepsilon < \varepsilon_c$. In this way we have

$$Z = \int \mathrm{d}E\,\mathcal{N}_{typ}(E)e^{-\beta E} = \int_{\varepsilon_c} \mathrm{d}\varepsilon\; e^{N[s(\varepsilon)-\beta\,\varepsilon]}. \tag{5.6}$$

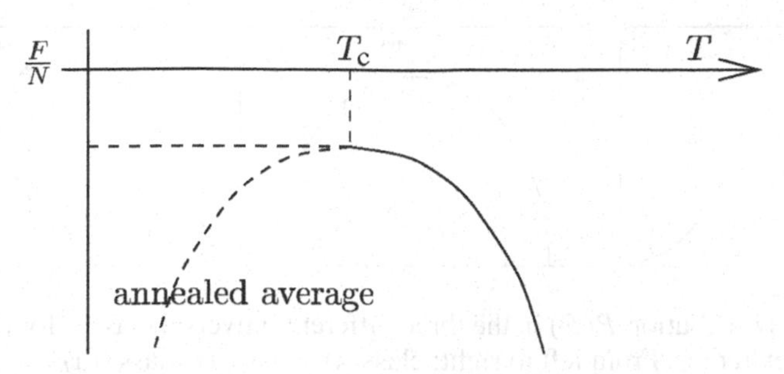

Figure 5.2 Free energy density as a function of the temperature. The dotted curve represents the annealed computation

The saddle point approximation leads to:

$$-2\varepsilon - \beta = 0 \longrightarrow \varepsilon_{sp} = -\frac{1}{2T}, \tag{5.7}$$

giving

$$F/N = -T \ln 2 - \frac{1}{4T}, \tag{5.8}$$

for $\varepsilon_{sp} \geq \varepsilon_c$. For $T < T_c$, ε_{sp} goes outside the integration range and the energy of the system remains stuck at ε_c. In Fig. 5.2 the dotted curve represents the analytical continuation we would get if considering the annealed result $\varepsilon(T) = 1/(2T)$ also for $T < T_c$. On the other hand, as we have seen, what really happens below T_c is that the levels with $\varepsilon = \varepsilon_c$ dominate. In terms of the thermodynamical weights $W_i = e^{-\beta E_i}/Z$, the only levels to have a finite weight are those with energy $E_i = N\varepsilon_c + u_i$ (with $u_i \sim O(1)$).

5.1.1 Statistics of extremes

A deeper comprehension of the system behaviour can be achieved by looking at the statistics of the dominant energy levels in the low temperature region, and in particular at the statistics of the extremes (Mézard, Parisi and Virasoro, 1985; Bouchaud and Mézard, 1997).

The distribution probability of the shift u_i from the critical value ε_c is easily computed. We have

$$P(u_i) = A\, e^{\frac{-(N\varepsilon_c + u_i)^2}{N}} \tag{5.9}$$

and, remembering that $\varepsilon_c = -\sqrt{\ln 2}$ and $T_c/2 = \sqrt{\ln 2}$, we get

$$P(u_i) \sim 2^{-N}\, e^{u_i/T_c}\left[1 + O\left(\frac{1}{N}\right)\right]. \tag{5.10}$$

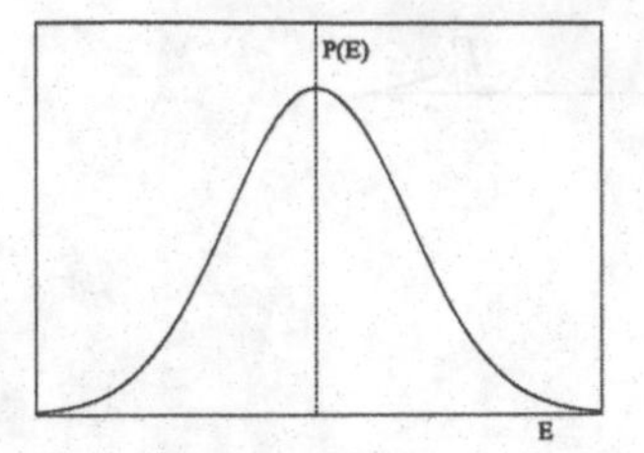

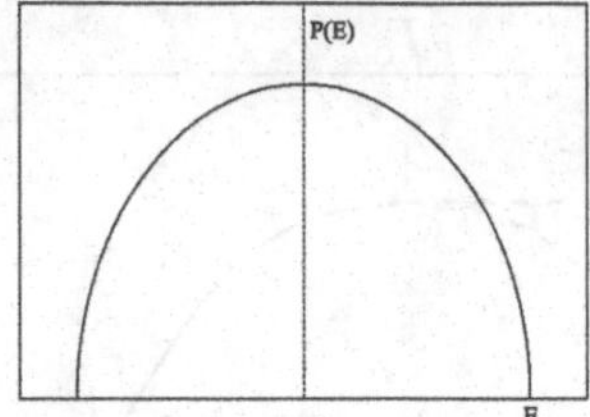

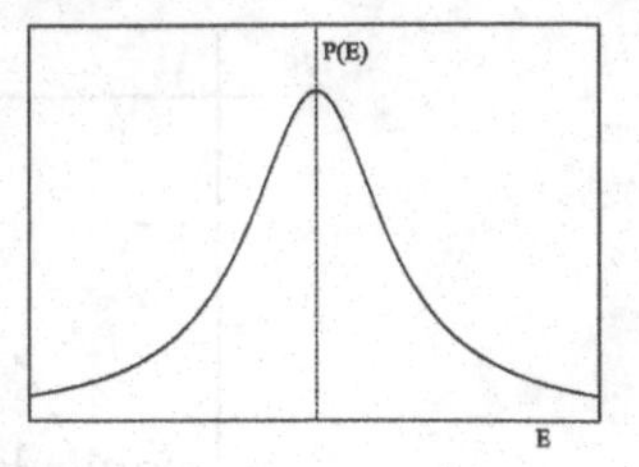

Figure 5.3 Distribution $P(E)$ in the three different universal classes for the statistics of the extremes. From left to right: class (i), class (ii), class (iii)

When looking at the energy axis this means that the density of levels increases with higher u_i.

Another important quantity is the distribution of the minimum. More precisely, given M levels $E_1, E_2, \ldots, E_M$, we can define $P^*(E)$ as the probability that the minimum among the M levels has energy E. We have:

$$P^*(E) = MP(E)\left[1 - \int_{-\infty}^{E} \mathrm{d}E'\, P\left(E'\right)\right]^M . \tag{5.11}$$

This can be better expressed in terms of the shift from the lower band edge. Introducing $E = N\varepsilon_c + u$, we find

$$P^*(u) \sim A\, \mathrm{e}^{u/T_c}\, \mathrm{e}^{-A\mathrm{e}^{u/T_c}} \tag{5.12}$$

and, introducing the re-scaled variable $\eta = Au/T_c$,

$$P^*(\eta) \sim \mathrm{e}^{\eta - \mathrm{e}^{\eta}} . \tag{5.13}$$

This is the so-called *Gumbel* distribution, which is a universal distribution class for the statistics of the extremes. Note that the maximum of $P^*(\eta)$ is at $\eta = 0$, meaning that the most probable value for the extreme energy (i.e. the ground state) is $u = 0$, i.e. $\varepsilon = \varepsilon_c$.

In general, when considering random *independent* variables E distributed according to $P(E)$, there are three different universality classes for the statistics of the extremes (Gumbel, 1958), according to the behaviour of the distribution in the tails (see Fig. 5.3):

(i) The first class where $P(E)$ decreases faster than any power (which is also the case of the REM) corresponds to the *Gumbel* distribution $P^*(u) \sim \mathrm{e}^{u - \mathrm{e}^u}$.
(ii) The second class where $P(E)$ is a bounded distribution corresponds to the *Weibull* bounded distribution.
(iii) Finally, the third class where $P(E)$ decreases as a power law and the statistics of the extremes is also described by a power law (*Frechet* distribution).

In the case of spin glasses, it is not a priori clear what happens, since energies are *not* in general uncorrelated variables. For example, in the SK model where a full RSB takes place and there are nontrivial correlations among the energies, the statistics of extremes does not belong to any of the three known universal classes (see Franz and Parisi, 2000).

From the distribution of the u_i (see Eq. (5.9)) we can also obtain the distribution of the weights for $T < T_c$. Indeed, we have

$$W_i = \frac{e^{-\beta E_i}}{Z} = \frac{e^{-\beta E_i}}{e^{-\beta E_i} + \sum_{j \neq i} e^{-\beta E_j}} = \frac{e^{-\beta E_i}}{e^{-\beta E_i} + Z_i'}, \tag{5.14}$$

and, consequently,

$$E_i = -\frac{1}{\beta} \ln\left(\frac{W_i Z'}{1 - W_i}\right). \tag{5.15}$$

This immediately gives

$$P(W_i = W)\, dW = P\left(E_i = -\frac{1}{\beta} \ln\left(\frac{WZ'}{1 - W}\right)\right) dE_i. \tag{5.16}$$

In the low temperature phase, as we have seen, W_i is finite only when $E_i = N\varepsilon_c + u_i$. The distribution of W is then easily expressed in terms of the distribution of the shift; using (5.9) we then find

$$\begin{aligned} P(W) &\sim e^{1/T_c\left(-T \ln\left(\frac{WZ'}{1-W}\right)\right)} \left(\frac{1}{W} + \frac{1}{1 - W}\right) \\ &= \left(\frac{WZ'}{1 - W}\right)^{-T/T_c} \frac{1}{W(1 - W)}. \end{aligned} \tag{5.17}$$

Introducing $\mu = \dfrac{T}{T_c} < 1$, we finally get

$$P(W) = \frac{W^{-1-\mu}(1 - W)^{-1+\mu}}{\Gamma(\mu)\Gamma(1 - \mu)}. \tag{5.18}$$

This result is very general and depends only on the statistics of extremes. From it we can deduce several properties of the low temperature phase. First, we note that Eq. (5.18) is not integrable, showing that the average multiplicity of levels is infinite: even if these levels are not exponentially large in number (since their entropy $s(\varepsilon_c)$ is zero), their number still grows as a power of the system size, diverging in the thermodynamic limit. Second, since $P(W)$ diverges at zero, there are an infinite number of levels with an infinitesimally small weight, while a finite number of levels have $O(1)$ weight. This can also be seen by looking at the moments of the distribution, $Y_k = \overline{\sum_i W_i^k}$. If all the $M = 2^N$ levels equally contribute to the thermodynamics one should have $W_i \sim O(1/M)$ for $i = 1, \ldots, M$ and $Y_k \sim M^{1-k} \to 0$ for $k > 1$. Using (5.18) we instead find that the moments (for $k > 1$) are *finite*. For example, for the second moment one has

$$\overline{W^2} = \sum_i \overline{W_i^2} = 1 - \mu = 1 - \frac{T}{T_c}, \tag{5.19}$$

indicating that there is a condensation of the weight on a few relevant states of low energy.

5.1.2 The REM via replicas

The REM can also be solved using the *replica approach* (Derrida, 1981; Bouchaud and Mézard, 1997). In this context we want to compute the average free energy $F = -1/\beta\, \overline{\ln Z}$. To do that we exploit the usual limit:

$$\overline{\ln Z} = \lim_{n \to 0} \frac{\overline{Z^n} - 1}{n},$$

which relates the computation of F to the one of $\overline{Z^n}$.

In the case of the REM $Z = \sum_i e^{-\beta E_i}$ and thus:

$$Z^n = \sum_{i_1,\ldots,i_n} \exp\{-\beta (E_{i_1} + \cdots + E_{i_n})\} = \sum_{i_1,\ldots,i_n} \exp\left\{-\beta \sum_j E_j \left(\sum_{a=1}^n \delta_{j;i_a}\right)\right\}. \tag{5.20}$$

After averaging over $P(E_i) \propto e^{-E_i^2/N}$ we get

$$\overline{Z^n} = \sum_{i_1,\ldots,i_n} \exp\left\{\frac{\beta^2 N}{4} \sum_j \left(\sum_{a=1}^n \delta_{j;i_a}\right)^2\right\} = \sum_{i_1,\ldots,i_n} \exp\left\{\frac{\beta^2 N}{4} \sum_{a,b} \delta_{i_a;i_b}\right\}. \tag{5.21}$$

When $N \to \infty$, one has to look at the saddle points of the effective action appearing in Eq. (5.21). Different ansatzes can be used for the symmetry properties of the saddle point.

- **RS solution**
  The simplest possibility is to consider a symmetric saddle point where *all the i_a are distinct*. We get from (5.21)

$$\overline{Z^n} = M(M-1)\cdots(M-n+1)\, e^{\frac{\beta^2}{4} Nn} = 2^{nN}\, e^{\frac{\beta^2}{4} nN}, \tag{5.22}$$

  which gives for the free energy

$$\frac{F}{N} = -T \ln 2 - \frac{1}{4T} \tag{5.23}$$

  (to be compared with Eq. (5.8)).

Note that the other symmetric saddle point $i_1 = i_2 = \cdots = i_n$, which gives $\overline{Z^n} \sim M \exp(\beta^2 n^2 N/4)$, is not acceptable since it leads to a value of $\overline{Z^n}$ which does not go to 1 when $n \to 0$.

This solution, as we have seen within the microcanonical approach, gives the correct result only for $T > T_c$. At lower temperatures it fails, on account of the contribution of rare atypical samples. Indeed, it is simple to verify that within the RS ansatz in zero external field $\overline{\ln Z} = \ln \overline{Z}$, reproducing the annealed average.

- **One-step RSB solution**

To recover the correct behaviour when $T < T_c$ one has to use a more complex ansatz that explicitly breaks the replica symmetry. We divide the n replicas in n/m distinct groups of m identical elements each, such that:

$$
\begin{aligned}
i_1 = &\quad i_2 = \cdots = i_m \\
i_{m+1} &\quad = \cdots \quad = i_{2m} \\
\vdots & \\
i_{n-m+1} &\quad = \cdots \quad = i_n
\end{aligned}
$$

In this way the replica symmetry is broken globally, but still holds within each sub-group. This is called a one-step breaking, and m is the *breaking parameter*. With this ansatz we find

$$
\begin{aligned}
\overline{Z^n} &= M(M-1)\cdots\left(M - \frac{n}{m} + 1\right)\frac{n!}{(m!)^{n/m}} \exp\left(\frac{\beta^2 N}{4}\frac{n}{m}m^2\right) \\
&\sim \exp\left(N\frac{n}{m}\ln 2 + Nnm\frac{\beta^2}{4}\right).
\end{aligned} \tag{5.24}
$$

This gives an expression of the free energy as a function of the breaking parameter m which can be fixed by a variational procedure. Extremization with respect to m leads to

$$
\begin{aligned}
-\frac{1}{m^2}\ln 2 + \frac{\beta^2}{4} &= 0, \\
m^2 &= 4\ln 2\, T^2,
\end{aligned} \tag{5.25}
$$

and, remembering that $T_c = 1/2\sqrt{\ln 2}$, we get

$$
m = \frac{T}{T_c}, \tag{5.26}
$$

which gives for the free energy for $T < T_c$:

$$
\frac{F}{N} = -\sqrt{\ln 2}. \tag{5.27}
$$

This RSB ansatz is then able to give the correct result. It, however, appears at this point as an artificial mathematical construction, without any physical justification. We shall discover in the next section, when addressing the low temperature phase of p-spin models, that actually this ansatz does correspond to a well defined thermodynamical scenario.

For the time being, let us focus on some formal features of this ansatz and note that the inequality $1 \leq m \leq n$, valid for n, m integers, becomes $n = 0 \leq m \leq 1$ in the limit $n \to 0$. The solution (5.26) is therefore in the range of admissible values for m (i.e. the interval [0, 1]) as long as $T < T_c$. On the other hand, for $T = T_c$, extremization of (5.24) leads to $m = 1$ and we recover the high temperature result.

5.2 The p-spin model

The p-spin model is defined by the following Hamiltonian.

$$\mathcal{H} = - \sum_{i_1 < i_2 < \cdots < i_p} J_{i_1 \cdots i_p} S_{i_1} \cdots S_{i_p}, \qquad (5.28)$$

where in the Ising version $S_i = \pm 1$, while in the spherical version the S_i are real variables satisfying a spherical constraint $\sum_i S_i^2 = N$. The $J_{i_1 \ldots i_p}$ are independent variables with distribution probability:

$$P(J) \sim \mathrm{e}^{-\frac{J^2 N^{p-1}}{p!}}; \qquad (5.29)$$

the variance $\overline{J^2}$ scales with N in such a way that the local field acting on one spin is of order 1, that is

$$\sum_{i_2,\ldots,i_p} J_{i_1 i_2 \ldots i_p} S_{i_2} \cdots S_{i_p} \sim N^{\frac{p-1}{2}} (\overline{J^2})^{1/2} \sim O(1).$$

The p-spin model is a mean field model. As we shall see, for $p \geq 3$ it belongs to the same class as the REM with a low temperature phase described by a one-step RSB ansatz. The case $p = 2$ is somewhat special. In the spherical case we recover the model studied in Chapter 3, namely a disguised ferromagnet which is solved by a simple RS solution. In the Ising case, on the other hand, we recover the Sherrington–Kirkpatrick model that has a more complex behaviour and will require a more complicated replica broken ansatz. In the following sections we consider p-spin with $p \geq 3$, and we focus on the Ising case. Analogous results are qualitatively valid also for the spherical version (see Crisanti and Sommers, 1992, and e.g., the reviews of Barrat, 1997; Giardina, 1998; Biroli, 1999; Castellani and Cavagna, 2005).

5.2.1 Relationship with the REM

Let us now see how the p-spin model is related to the Random Energy Model studied in the previous sections. The REM is defined via the Gaussian distribution of $M = 2^N$ independent energy levels. In the present case each configuration $\{S_i\}$ corresponds to an energy $\mathcal{H}(S)$ that fluctuates due to the presence of the disordered couplings. The probability that the configuration $\{S_i\}$ has an energy E is then

$$\begin{aligned} P(\mathcal{H}(S)=E) &= \int \mathrm{d}J_{i_1\cdots i_p}\, \mathrm{e}^{-J^2_{i_1\cdots i_p}\frac{N^{p-1}}{p!}}\,\delta(H(S)-E) \\ &= \prod_{i_1<\cdots<i_p} \int \frac{\mathrm{d}\lambda}{2\pi}\int \mathrm{d}J_{i_1\ldots i_p}\, \mathrm{e}^{\mathrm{i}\lambda E+\mathrm{i}\lambda J_{i_1\ldots i_p}S_{i_1}\ldots S_{i_p}-J^2_{i_1\ldots i_p}\frac{N^{p-1}}{p!}} \\ &= \int \frac{\mathrm{d}\lambda}{2\pi}\,\mathrm{e}^{\mathrm{i}\lambda E}\left[\mathrm{e}^{-\frac{\lambda^2 p!}{4N^{p-1}}}\right]^{\frac{N^p}{p!}} \sim \mathrm{e}^{-\mathrm{E}^2/N}. \end{aligned} \tag{5.30}$$

Thus we have

$$P(\mathcal{H}(S)=E) \sim \mathrm{e}^{-\frac{E^2}{N}} \tag{5.31}$$

as in the REM. There is, however, an important difference: in the REM the energy levels are *independent* variables. This is not the case here, since different configurations may be correlated to different degrees. Indeed, let us consider two configurations $\{S_i\}$, $\{S_i'\}$ with overlap $\frac{1}{N}\sum_i S_i S_i' = Q$. The probability that $\mathcal{H}(S)=E$ and $\mathcal{H}(S')=E'$ is easily computed as

$$P(\mathcal{H}(S)=E,\mathcal{H}(S')=E')|_Q \sim \mathrm{e}^{-\frac{(E+E')^2}{2N(1+Q^p)}-\frac{(E-E')^2}{2N(1-Q^p)}}. \tag{5.32}$$

Thus, in general energy levels are correlated. However, when the overlap is $Q<1$, in the limit $p\to\infty$ we find:

$$P(E,E') \underset{p\to\infty}{\to} \mathrm{e}^{-\frac{E^2}{N}}\,\mathrm{e}^{-\frac{E'^2}{N}}, \tag{5.33}$$

that is the energies become uncorrelated variables, as in the REM model.

Note that setting $Q<1$ before the limit is crucial: if we consider two configurations $\{S_i\}$ and $\{S_i'\}$ which differ at a finite number of sites (i.e. $Q=1$), they remain correlated even in the $p\to\infty$ limit.

5.2.2 The replica approach

The statics of the Ising p-spin Model can be addressed with the replica method (Gross and Mézard, 1984; Gardner, 1985). Again, we want to compute $F = -(1/\beta)\,\overline{\ln Z}$ using the replica trick $\overline{\ln Z} = \lim_{n\to 0}(\overline{Z^n}-1)/n$. We have;

$$Z = \sum_{\{S_i\}} \mathrm{e}^{\beta \sum\limits_{i_1<\cdots<i_p} J_{i_1\cdots i_p}S_{i_1}\cdots S_{i_p}}. \tag{5.34}$$

We replicate introducing n replicas $S_i^a (i = 1, \ldots, N, a = 1, \ldots, n)$. After averaging over the disorder $J_{i_1 \ldots i_p}$, we get

$$\overline{Z^n} = \sum_{\{S_i^a\}} \exp\left\{ \frac{\beta^2}{4} \frac{p!}{N^{(p-1)}} \sum_{i_1 < \cdots < i_p} \sum_{a,b} S_{i_1}^a S_{i_1}^b \, S_{i_2}^a S_{i_2}^b \cdots S_{i_p}^a S_{i_p}^b \right\}. \tag{5.35}$$

We introduce the *overlap* variables $Q_{ab} = \frac{1}{N} \sum_i S_i^a S_i^b$ using the identity

$$1 = \prod_{a<b} \int \mathrm{d}Q_{ab} \int \frac{\mathrm{d}\lambda_{ab}}{2\pi} \mathrm{e}^{-\lambda_{ab}\left(N Q_{ab} - \sum_i S_i^a S_i^b\right)}, \tag{5.36}$$

where λ_{ab} runs along the imaginary axis. Then we get

$$\overline{Z^n} = \int \prod_{a<b} \left(\mathrm{d}Q_{ab} \frac{\mathrm{d}\lambda_{ab}}{2\pi} \right) \mathrm{e}^{-NG(Q,\lambda)}, \tag{5.37}$$

where the effective action G, within subextensive terms, is given by:

$$G(Q, \lambda) = -\frac{\beta^2}{4} \sum_{a,b} Q_{ab}^p + \frac{1}{2} \sum_{a<b} \lambda_{ab} Q_{ab} - \ln \operatorname*{tr}_{\{S\}} \left(\mathrm{e}^{\frac{1}{2} \sum_{a \neq b} \lambda_{ab} S_a S_b} \right). \tag{5.38}$$

Both λ_{ab} and Q_{ab} are $n \times n$ symmetric matrices. Note that the first sum in (5.38) includes a term with $a = b$. Indeed, the term with $Q_{aa} = (1/N) \sum_i S_i^a S_i^a = 1$ is present in the replicated action and must be taken into account. Q_{aa} is not, however, an integration variable, its value being fixed to one by definition. The action G thus depends on $n(n-1)$ variables.

The saddle point of the action is given by

$$\frac{\partial G(Q, \lambda)}{\partial Q_{ab}} = 0; \qquad \frac{\partial G(Q, \lambda)}{\partial \lambda_{ab}} = 0. \tag{5.39}$$

5.2.3 The RS solution

The action G is invariant with respect to permutations of replica indices (i.e. the symmetry group of G is P_n), the simplest possibility is that the saddle point has the same symmetry as the action. We therefore assume

$$Q_{ab} = Q; \qquad \lambda_{ab} = \lambda, \quad \text{for } a \neq b. \tag{5.40}$$

The saddle point Eqs. (5.39) then give

$$\begin{cases} \lambda = \dfrac{\beta^2}{2} p \, Q^{p-1}, \\ Q = \displaystyle\int \frac{\mathrm{d}z}{\sqrt{2\pi}} \mathrm{e}^{-\frac{z^2}{2}} \tanh(z\sqrt{\lambda}). \end{cases}$$

In the limit $p \to \infty$ they are easy to solve, yielding

$$Q = 0; \qquad \lambda = 0, \tag{5.41}$$

which gives for the free energy

$$\frac{F}{N} = -\frac{1}{4T} - T \ln 2, \tag{5.42}$$

as in the REM model. We know that for the REM this solution is correct for $T > T_c$, but is not valid for $T < T_c$. This indicates that for the p-spin at low T we also have to look for other saddle points which do not have the symmetry of the action.

5.2.4 Overlap distribution and Replica Symmetry Breaking

We have learnt in the case of the REM that in the low temperature phase the weight is concentrated on a finite number of levels with energy close to the minimal one and with zero entropy. Given the analogy between the two models, we expect the same to be true for the p-spin. In this case, however, we do not deal with energy levels but configurations that live in a well defined phase space. At equilibrium some configurations, or *microstates*, are weighted more than others, exactly as in the REM. We may ask how these relevant microstates are organized in the phase space. For example, we may ask what is the distance between equilibrium configurations at a given temperature. The quantity which describes this feature is the $P_J(Q)$, defined as the probability distribution that two configurations have at equilibrium mutual overlap (i.e. co-distance) given by Q. Its explicit definition is then

$$P_J(Q) = \sum_C \frac{e^{-\beta E_C}}{Z} \sum_{C'} \frac{e^{-\beta E_{C'}}}{Z} \delta(Q^{CC'} - Q), \tag{5.43}$$

where

$$Q^{CC'} = \frac{1}{N} \sum_i S_i S_i' \qquad \text{and} \qquad C \equiv \{S_i\},\ C' \equiv \{S_i'\}. \tag{5.44}$$

From the shape of the overlap distribution we can gather important information about the thermodynamics of the system. If only one pure state (i.e. ergodic component) is present, as is the case at high temperature, two equilibrium configurations are typically sampled at a mutual overlap equal to the self-overlap of the state (the self-overlap going to one if the state were to comprise a single configuration). The $P_J(Q)$ then is a delta function at the self-overlap (zero for the paramagnetic state at high temperature). However, if different pure states are present, a nontrivial structure for the distribution is expected. A simple example is given by the ferromagnet: at low temperatures two equilibrium configurations may belong to the same state with magnetization m, and their overlap is then given by $Q = m^2$, or to different states, corresponding to $Q = -m^2$. The overlap distribution in this case displays two delta functions.

In general, the support of the $P_J(Q)$ corresponds to all the possible mutual overlaps among the pure (macro) states present in the system. To better see this, let us assume that different pure states exist and label each one of them with an index α. The equilibrium measure $\langle \cdots \rangle = \sum_C \exp\{-\beta E_C\}/Z(\cdots)$ then decomposes as $\langle \cdots \rangle = \sum_\alpha W_\alpha \langle \cdots \rangle_\alpha$, where $W_\alpha = Z_\alpha / Z$ indicates the thermodynamic weight of the α state. Using this decomposition, Eq. (5.43) can then be written as

$$P_J(Q) = \sum_{\alpha,\beta} W_\alpha W_\beta \, \delta(Q^{\alpha\beta} - Q). \tag{5.45}$$

The equivalence between (5.43) and (5.45) relies on the clustering property of pure states, i.e. $\langle S_i S_j \rangle_\alpha \sim \langle S_i \rangle_\alpha \langle S_j \rangle_\alpha$ in the thermodynamic limit. Due to this, all the moments $\langle Q^k \rangle$ computed with the two definitions are easily shown to be the same (Mézard, Parisi and Virasoro, 1987).

From Expression (5.45) we clearly see that a nontrivial support for the overlap distribution indicates a nontrivial structure of states. Note also that, in all the following computations, we consider only *positive* values of the overlap (i.e. $Q \in [0, 1]$). In this way all the information provided by the overlap distribution concerns pure states *not* related by the inversion symmetry, contrary to the ferromagnetic example.

In a spin glass, $P_J(Q)$ depends on the quenched disorder J_{ij} through E_C, $E_{C'}$ and Z (or, if expressed as in (5.45), through the weights W_α, W_β). The simplest thing to do is to compute the average distribution

$$P(Q) = \overline{P_J(Q)}. \tag{5.46}$$

To this end, one can resort once again to replicas (Parisi, 1983). Using the expression

$$\frac{1}{Z^2} = \lim_{n \to 0} Z^{n-2}, \tag{5.47}$$

we get

$$P(Q) = \lim_{n \to 0} \sum_{C_1,\ldots,C_n} \overline{\mathrm{e}^{-\beta \sum_a E_{C_a}}} \, \delta\left(Q^{C_1 C_2} - Q\right). \tag{5.48}$$

The calculation is very similar to the one of $\overline{Z^n}$, the difference is that now the overlap between replicas 1 and 2 is fixed to be Q by the delta function. We therefore get

$$P(Q) = \langle \delta(Q_{12} - Q) \rangle_G, \tag{5.49}$$

where $\langle \ldots \rangle_G$ indicates an average with respect to the measure given by the action G of Eq. (5.38). When performing the saddle point approximation one has that

$$P(Q) = \delta(Q_{12} - Q)|_{\text{SP of } G}. \tag{5.50}$$

In the case of the RS solution we then get

$$P(Q) = \delta(Q) \tag{5.51}$$

with only one possible null overlap for equilibrium configurations, suggesting that only one pure state, the paramagnetic one, is present (see, however, Section 5.2.6 for another interpretation).

What is happening when we consider saddle points which break the replica symmetry? Imagine that we find a saddle point $Q_{\text{SP}} = \{Q_{ab}\}$ that breaks the replica permutation symmetry; then because of the symmetry $(Q^\pi)_{ab} = Q_{\pi(a),\pi(b)}$ (with $\pi \in S_n$) is also a saddle point having the same action. When computing $P(Q)$ we have therefore to sum up over all equivalent saddle points:

$$P(Q) = \sum_\pi \delta\left(Q^\pi_{12} - Q\right)\Big|_{\text{SP}}, \tag{5.52}$$

which gives

$$P(Q) = \frac{1}{n(n-1)} \sum_{a \neq b} \delta\left(Q^{ab}_{\text{SP}} - Q\right). \tag{5.53}$$

The definition of the overlap distribution $P(Q)$ and its expression (5.53) give us a hint for selecting RSB good saddle points. Indeed, we must require that the saddle point gives for $P(Q)$ a sensible result, that is $P(Q)$ should satisfy the properties of a probability distribution.

5.2.5 The one-step RSB solution

It turns out that the correct ansatz for the overlap matrix is very similar to the one adopted for the REM. Let us divide the $n \times n$ symmetric matrix Q into sub-blocks in the following way. We choose n/m blocks of size m along the diagonal. We set $Q_{ab} = Q_1$ inside the diagonal blocks (the diagonal itself fixed to one, since $S_i^a = \pm 1$), and $Q_{ab} = Q_0$ outside (see Fig. 5.4). Namely,

$$Q_{ab} = (1 - Q_1)\delta_{a;b} + (Q_1 - Q_0)\varepsilon_{a;b} + Q_0, \tag{5.54}$$

where $\varepsilon_{a;b}$ is set to be equal to one if the two replica indices a and b belong to the same block and zero otherwise.

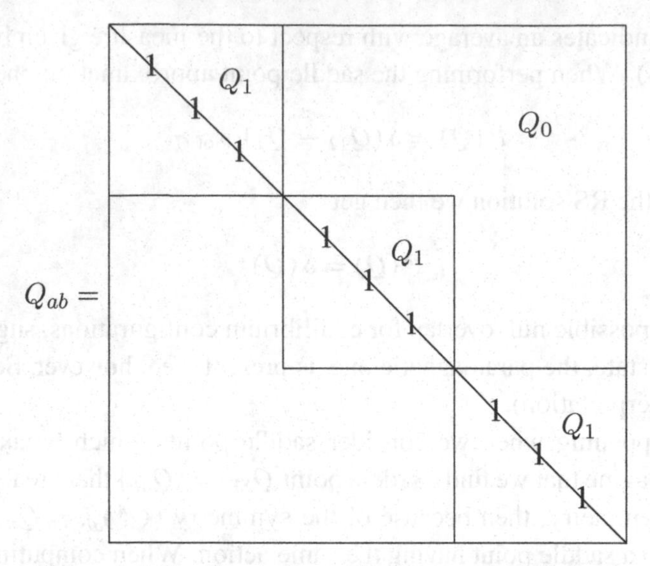

Figure 5.4 Overlap matrix with a one-step RSB ansatz

With this ansatz for Q_{ab} we get

$$
\begin{aligned}
P(Q) &= \frac{1}{n(n-1)} n\left[(n-m)\,\delta(Q-Q_0) + (m-1)\,\delta(Q-Q_1)\right] \\
&= \frac{1-m}{1-n}\delta(Q-Q_1) + \frac{m-n}{1-n}\delta(Q-Q_0).
\end{aligned}
\tag{5.55}
$$

When n, m are positive integers we have

$$1 \le m \le n.$$

However, the requirement that $P(Q)$ is positive implies that in the limit $n \to 0$, the inequality has its order becoming: $n \le m \le 1 \Longrightarrow 0 \le m \le 1$. In this way we have

$$P(Q) = (1-m)\,\delta(Q-Q_1) + m\delta(Q-Q_0), \tag{5.56}$$

where Q_1, Q_0 and m have to be fixed by a variational procedure (see below).

The computation of the free energy is also straightforward. We have to use the ansatz (5.54) both for the overlap matrix Q_{ab} and for the matrix of the auxiliary variables λ_{ab} appearing in (5.38). For example, using (5.54) the term $\sum_{a\neq b} Q_{ab}^p$ becomes $n[(m-1)Q_1^p + (n-m)Q_0^p]$. In general, the action becomes a variational function of a limited number of parameters. In this case

$$G(Q_{ab}, \lambda_{ab}) \to G(Q_1, Q_0, \lambda_1, \lambda_0, m). \tag{5.57}$$

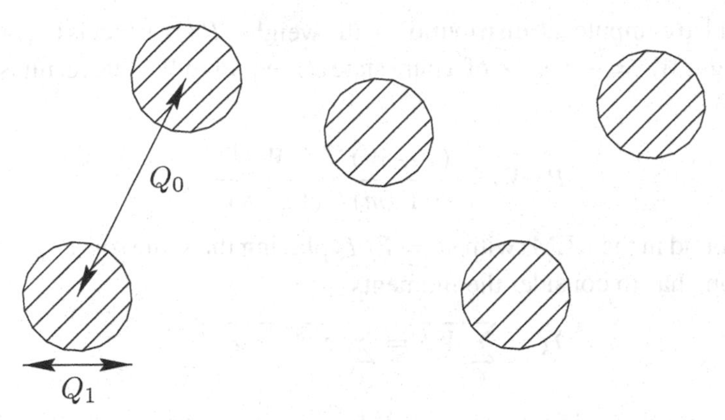

Figure 5.5 Pictorial representation of the structure of states for the p-spin at low temperature

Then the action has to be extremized with respect to Q_1, Q_0, λ_1, λ_0 and with respect to the breaking parameter m. Note that the multiplicities of Q_1 and Q_0 (respectively $1-m$ and $m-n$) change sign in the $n \to 0$ limit, which entails that the action convexity changes sign as well, leading to a maximum of the free energy with respect to those variables, instead of a minimum as in the free energy of pure systems.

One finds a saddle point with $m \leq 1$, i.e. compatible with the interpretation of $P(Q)$ as a probability, only for $T \leq T_c$. In the limit $p \to \infty$ this saddle point gives

$$\begin{cases} Q_0 = 0, \\ Q_1 = 1, \\ m = T/T_c, \end{cases} \tag{5.58}$$

and we get back the REM. For finite p (and $T > 0$) on the other hand one gets $0 < Q_1 < 1$, and $Q_0 = 0$.

Let us now look back at the $P(Q)$ to see how this RSB solution can be interpreted in terms of structure of pure states. We find

$$P(Q) = m\delta(Q) + (1-m)\,\delta(Q - Q_1)\,. \tag{5.59}$$

This form of the $P(Q)$ means that only two values of the overlap are statistically significant at equilibrium. Then, in the phase space, the configurations 'gather' into distinct domains (the pure states), each one having a self-overlap $Q_1 < 1$, and with typical mutual overlap given by $Q_0 = 0$ (see Fig. 5.5). The REM is a limit where each pocket-domain shrinks to one point, i.e. to a single configuration (indeed for $p \to \infty$, $Q_1 \to 1$). The above results can also be derived via the cavity method, and with a rigorous mathematical approach (Talagrand, 2000).

One can also compute the distribution of the weights $P(W)$, that is the probability that the weight $W_\alpha = -\beta F_\alpha / Z$ of a pure state α is equal to W. The result is (Mézard *et al.*, 1984)

$$P(W) = \frac{(1-W)^{-1-m}\, W^{-1+m}}{\Gamma(m)\Gamma(1-m)}, \tag{5.60}$$

exactly as found in the REM (with $m = T/T_c$ playing the same role as μ). To obtain this result one has to consider the moments

$$Y_k = \sum_C \overline{W_c^k} = \sum_C \overline{e^{-\beta k E_c} Z^{-k}}. \tag{5.61}$$

Again, one can use replicas to express $Z^{-k} = \lim_{n\to 0} Z^{n-k}$ and one finally gets:

$$Y_k = \sum_{C_1,\ldots,C_n} \delta\left(Q^{C_1C_2} - 1\right) \cdots \delta\left(Q^{C_1C_k} - 1\right) \delta\left(Q^{C_1C_{k+1}}\right) \cdots, \tag{5.62}$$

where the Q matrix has the form specified in the one-step ansatz. In (5.62) only the terms where the first k replicas belong to the same block are different from zero. Remembering that the size of each block is of m replicas, we have to compute the number of ways of putting k replicas in the same block of size m. It is given by

$$\frac{n(m-1)\cdots(m-k+1)}{n(n-1)\ldots(n-k+1)}, \tag{5.63}$$

which, for $n \to 0$, is $(1-m)(2-m)\cdots(k-1-m)/(k-1)!$. Thus we get

$$Y_k = \frac{\Gamma(k-m)}{\Gamma(k)\Gamma(1-m)}. \tag{5.64}$$

But these are precisely the moments one gets from the distribution $P(W)$ as in Eq. (5.18), showing that it is indeed the same weight distribution.

5.2.6 Metastability, complexity and glassy behaviour

We have seen for the REM that a microcanonical approach can be implemented to compute the thermodynamics which unveils the transition as a condensation on energy levels associated with zero entropy. Something similar can actually be done also for the p-spin model.

A more detailed analysis of this system using mean field equations for the local magnetizations (the so-called TAP approach discussed in Chapter 7), reveals that the p-spin has a huge number of metastable states, i.e. states that are locally stable (and thus correspond to stable solutions of the mean field equations) but with free energy density higher than the ground state. In particular, one can show that the number $\mathcal{N}(f)$ of states with free energy density f scales exponentially with the size of the system, therefore defining a new entropy

$$\Sigma(f) = \frac{1}{N}\overline{\ln \mathcal{N}(f)}, \tag{5.65}$$

usually called the *complexity*. $\Sigma(f)$ has been explicitly computed both for the Ising (Rieger, 1992) and for the spherical (Crisanti and Sommers, 1995; Crisanti, Leuzzi and Rizzo, 2003) versions. It is an increasing function of f defined on a finite interval $[f_0, f_{\text{th}}]$, with $\Sigma(f_0) = 0$ and $\Sigma(f_{\text{th}}) = \Sigma_{\max}$. Then f_0 is the free energy of the ground states, which are the lesser in number, while f_{th} is the free energy of the threshold states with maximal entropy.

Given the complexity, we can easily compute the partition function. Indeed, we have:

$$Z = \int \mathrm{d}f \; \mathrm{e}^{-\beta N[f - T\Sigma(f)]} + Z_{\text{para}}, \tag{5.66}$$

where Z_{para} stands for the contribution of the paramagnetic state $m_i = 0$. Note that this is the analogue of Eq. (5.2), where, now, metastable states play the role of the energy levels in the REM.

In the thermodynamic limit one can use the saddle point approximation to evaluate the integral in (5.66). What is found is the following.

For high temperatures $T > T_{\text{d}}$ the saddle point f^* lies outside the admissible range $[f_0, f_{\text{th}}]$, i.e. $f^* > f_{\text{th}}$; the integral is then dominated by the threshold states, but anyway the contribution of Z_{para} is the leading one and the system is therefore found in the paramagnetic state.

For temperatures in the intermediate range $T_{\text{c}} < T < T_{\text{d}}$, the saddle point f^* lies in the interval $[f_0, f_{\text{th}}]$, and the partition function is given by $Z = \exp(-\beta F)$ with $F = f^* - T\Sigma(f^*)$. In this case the equilibrium is given by an exponentially large set of metastable states. Interestingly, since the typical overlap between any two of these states is zero, and the probability of picking the same state twice is exponentially small (their complexity being finite), the corresponding $P(Q)$ is a delta function at zero. In other words, what appears as a paramagnetic state within the replica computation is actually, below T_{d}, an exponentially large ensemble of metastable states (Cavagna, Giardina and Parisi, 1997). Note that T_{d} turns out also to be the temperature where the dynamical anomalies occur in this model. This is not a coincidence, but, again, a consequence of the deep relationship between dynamics and geometry of the phase space (see Cugliandolo and Kurchan, 1993; Cavagna, Garrahan and Giardina, 2000, for more discussion of this point).

As the temperature approaches T_{c} the saddle point f^* reaches the lower band edge f_0 and, for $T < T_{\text{c}}$, the integral is dominated by the ground states. Then, for the p-spin (as for the REM) the static transition occurs in correspondence to a condensation on a set of low (free) energy and zero entropy states. The difference with the REM is that now the entropy which goes to zero is the one associated with the number of states (while intra-state entropic contributions remain finite for $p < \infty$ and $T > 0$).

As first observed by Kirkpatrick and Thirumalai (1987), the qualitative scenario described above is in striking correspondence with the behaviour observed in supercooled glass-forming liquids (see, e.g., Kob, 2003). When a liquid is cooled below the melting point without undergoing the transition to the crystal state, it remains in a metastable liquid state, referred to as 'supercooled'. It has been speculated and shown numerically that this liquid state corresponds to a set of an exponentially large number of valleys sampled by the system in configuration space. Valleys are somehow the finite dimensional counterpart of the metastable mean field pure states of the p-spin model. In the supercooled liquid case also, the energy and complexity of these valleys decrease with temperature. In this context the freezing temperature T_c corresponds to the so-called Kauzmann temperature where an entropy crisis would occur. This temperature is, however, not attainable experimentally, since the viscosity becomes too high and relaxation to equilibrium practically impossible: these systems enter at low temperature an off-equilibrium glassy phase.

5.3 Summary

In this chapter we have analyzed two models which exhibit a one-step pattern of replica symmmetry breaking: the REM model and the p-spin model.

In the case of the REM the thermodynamics is also solved without using the replica method. This has helped us in understanding the mechanism of the freezing transition and checking the correctness of the broken replica ansatz valid at low temperatures. This simpler model has also offered us a precious guideline for the p-spin model (which reproduces the REM in the $p \to \infty$ limit), showing the way for the correct replica breaking.

What we understood of the physics of the one-step RSB models can be summarized by the following points.

- These systems have a huge number of metastable states (energy levels in the REM and pure states for the p-spin), with nonzero entropy in a finite (free) energy density interval.
- In the high temperature phase, states with finite entropy dominate, that have energy density higher than the ground state one. The mutual overlap between these states is $Q = 0$, so that their overlap distribution is given by $P(Q) = \delta(Q)$. Within the replica formalism they are described by a replica symmetric ansatz, and their global free energy contribution is equal to the paramagnetic free energy (even if, as we have seen, in the p-spin a true ergodic paramagnetic state is present only for $T > T_d$).
- As the temperature is lowered, the energy density and the entropy of the dominant equilibrium states decrease, and at the static transition temperature T_c a condensation occurs

on the zero entropy ground states. These states have, again, zero mutual overlap. However, being much less numerous than all the other ones, the probability that two randomly chosen equilibrium configurations belong to the same state is finite. Thus the overlap distribution $P(Q)$ now exhibits two delta functions in correspondence to the mutual overlap *and* the self-overlap. These states are described by a one-step RSB solution where the same two overlaps appear, together with the breaking parameter m (μ in the REM). As we have seen, the physical meaning of this parameter is related to the distribution of the weights of the ground states, i.e. m determines how the extensive (free) energy of these states is distributed around the lower band edge.

- There is a close connection with the statistics of the extremes, and we have seen that these models do belong to the Gumbel universal class.

References

Barrat A. (1997). Ph.D. Thesis, Preprint cond-mat/9701031.
Biroli G. (1999). Ph.D. Thesis, Preprint cond-mat/9909415.
Bouchaud J.-P. and Mézard M. (1997). *J. Phys. A*, **30**, 7997.
Castellani T. and Cavagna A. (2005). *J. Stat. Mech.*, P05012.
Cavagna A., Garrahan J. P. and Giardina I. (2000). *Phys. Rev. B*, **61**, 3960.
Cavagna A., Giardina I. and Parisi G. (1997). *J. Phys. A*, **30**, 4449 and 7021.
Crisanti A. and Sommers H. J. (1992). *Z. Phys. B*, **5**, 805.
Crisanti A. and Sommers H. J. (1995). *J. Phys. I (France)*, **87**, 341.
Crisanti A., Leuzzi L. and Rizzo T. (2003). *Eur. Phys. J. B*, **36**, 129.
Cugliandolo L. F. and Kurchan J. (1993). *Phys. Rev. Lett.*, **71**, 173.
Derrida B. (1981). *Phys. Rev. B*, **24,** 2613.
Franz S. and Parisi G. (2000). *Eur. Phys. J. B*, **18**, 485.
Gardner E. (1985). *Nucl. Phys. B*, **257**, 747.
Giardina I. (1998). Ph.D. Thesis, chimera.roma1.infn.it.
Gross D. and Mézard M. (1984). *Nucl. Phys. B*, **240**, 431.
Gumbel E. J. (1958). *Statistics of Extremes*, New York, Columbia University Press.
Kirkpatrick T. R. and Thirumalai D. (1987). *Phys. Rev. B*, **36**, 5388.
Kob W. (2003). Supercooled liquids, the glass transition, and computer simulations. In *Slow Relaxations and Nonequilibrium Dynamics in Condensed Matter*, Les Houches Session LXXVII, eds. J.-L. Barrat, M. Feigelman, J. Kurchan and J. Dalibard, Les Ulis, EDP Sciences/Springer-Verlag.
Mézard M. and Parisi G. (2000). *J. Phys: Condensed Matter*, **12**, 6655.
Mézard M., Parisi G. and Virasoro M. A. (1985). *J. Phys. Lett.*, **46**, L217.
Mézard M., Parisi G. and Virasoro M. A. (1987). *Spin Glass Theory and Beyond*, Singapore, World Scientific, Vol. 9.
Mézard M., Parisi G., Sourlas N., Toulouse G. and Virasoro M. A. (1984). *J. Physique*, **45**, 843.
Parisi G. (1979). *Phys. Rev. Lett.*, **43**, 1754.
Parisi G. (1983). *Phys. Rev. Lett.*, **50**, 1946.
Rieger H. (1992). *Phys. Rev. B*, **46**, 14655.
Sherrington D. and Kirkpatrick S. (1975). *Phys. Rev. Lett.*, **35**, 1792.
Talagrand M. (2000). *Cr. Acad. Sci. I-Math*, **331**, 939.

6

The Sherrington–Kirkpatrick Model

The Sherrington–Kirkpatrick Model (Sherrington and Kirkpatrick, 1975), or SK model as it is commonly referred to, is the long range version of the Edwards–Anderson Model (Edwards and Anderson, 1975) introduced to describe spin glass systems at low temperature. As we anticipated in the introduction of the previous chapter, this model has been fully solved within the replica method by Parisi in 1979, unveiling a new 'spin glass' phase whose nontrivial features are formally encoded in the ultrametric replica-symmetry broken solution. In this chapter we analyze in detail the SK model and its solution. We start in Section 6.2 by describing the replica symmetric solution, originally proposed by Sherrington and Kirkpatrick in their work. Then, in Section 6.3, we examine the stability of this solution using the method of the Replica Fourier Transform. In Section 6.4 we consider the broken solution, starting with a finite number of breaking steps and then taking the continuous limit. Finally, we give a brief interpretation of the mathematical results in terms of physical quantities and describe the hierarchical structure of states in the low temperature phase. A very useful book where most of the results discussed in this chapter are exhaustively treated is *Spin Glass Theory and Beyond* by Mézard, Parisi and Virasoro (1987).

6.1 The model

The SK model is defined by the following Hamiltonian:

$$\mathcal{H} = -\sum_{(ij)} J_{ij}\, S_i\, S_j - h \sum_i S_i, \tag{6.1}$$

where the spins S_i, $i = 1, \ldots, N$ are Ising variables and the sum runs over distinct pairs. The couplings J_{ij}, acting between all pairs, are quenched random variables with a probability law

$$\mathcal{P}(J_{ij}) \sim \mathrm{e}^{-N J_{ij}^2 / 2J^2}. \tag{6.2}$$

To compute the partition function we use, again, the basic relation of the replica method, namely $\ln Z = \lim_{n\to 0}[Z^n - 1]/n$. In this way, instead of dealing with the disorder average of the logarithm, we focus on the partition function Z^n of a replicated system of n distinct copies of the original one. From Eq. (6.1) we have:

$$Z^n = \operatorname*{tr}_{\{S_i^a\}} \exp\left\{\beta \sum_a \sum_{(ij)} J_{ij}\, S_i^a\, S_j^a + \beta h \sum_{ia} S_i^a\right\}, \tag{6.3}$$

and, performing the average over J_{ij},

$$\overline{Z^n} = \operatorname*{tr}_{\{S_i^a\}} \exp\left\{\frac{(\beta J)^2}{2N} \sum_{(ij)} \left(\sum_a S_i^a\, S_j^a\right)^2 + \beta h \sum_{ia} S_i^a\right\}. \tag{6.4}$$

By reordering the sums, we have

$$\begin{aligned}\frac{1}{N}\sum_{(ij)}\sum_{a,b} S_i^a\, S_j^a\, S_i^b\, S_j^b &= \frac{1}{N}\sum_{i,j}\sum_{(ab)} S_i^a\, S_i^b\, S_j^a\, S_j^b + \frac{Nn}{2} - \frac{n^2}{2} \\ &= N\sum_{(ab)}\left(\frac{1}{N}\sum_i S_i^a\, S_i^b\right)^2 + \frac{Nn}{2} + \mathrm{O}(n^2). \end{aligned} \tag{6.5}$$

Thus,

$$\overline{Z^n} = \operatorname*{tr}_{\{S_i^a\}} \exp\left\{\frac{N(\beta J)^2}{2}\sum_{(ab)}\left(\frac{1}{N}\sum_i S_i^a\, S_i^b\right)^2 + \beta h \sum_{ia} S_i^a + \frac{Nn(\beta J)^2}{4}\right\}. \tag{6.6}$$

To linearize the quartic term in (6.6) we use the Hubbard–Stratonovich transformation and introduce a set of overlap variables Q^{ab} (as in (4.82)):

$$\begin{aligned}\int \frac{\mathrm{d}Q^{ab}}{\sqrt{2\pi}}\sqrt{N(\beta J)^2}\exp\left\{-\frac{N(\beta J)^2}{2}(Q^{ab})^2 + (\beta J)^2 Q^{ab}\sum_i S_i^a\, S_i^b\right\} \\ = \exp\left\{\frac{(\beta J)^2}{2N}\left(\sum_i S_i^a\, S_i^b\right)^2\right\}.\end{aligned} \tag{6.7}$$

In this way we get

$$\overline{Z^n} = \int \prod_{(ab)}\left(\frac{\mathrm{d}Q^{ab}}{\sqrt{2\pi}}\right) \mathrm{e}^{\mathcal{L}\{Q^{ab}\}}, \tag{6.8}$$

where the effective Lagrangian $\mathcal{L}$, within subextensive terms, is given by[†]

$$\mathcal{L}\{Q^{ab}\} = -\frac{N}{2}(\beta J)^2 \sum_{(ab)} (Q^{ab})^2 + Nn\frac{(\beta J)^2}{4}$$
$$+ N \ln \operatorname*{tr}_{\{S^a\}} \exp\left\{ \frac{(\beta J)^2}{2} \sum_{ab} Q^{ab} S^a S^b + \beta h \sum_a S^a \right\}. \quad (6.9)$$

Here S^a is a single site variable and we have set by definition

$$Q^{ab} = Q^{ba}, \qquad Q^{aa} = 0. \quad (6.10)$$

Now, with N a factor we can use the saddle point method. We find

$$\frac{\partial \mathcal{L}}{\partial Q^{ab}} = 0 \quad \longrightarrow \quad Q^{ab} = \langle S^a S^b \rangle, \quad (6.11)$$

where the expectation value $\langle \cdot \rangle$ is calculated with the single site normalized weight $\zeta(S)/\mathrm{tr}\, \zeta(S)$, where

$$\zeta(S) = \exp\left\{ \frac{(\beta J)^2}{2} \sum_{ab} Q^{ab} S^a S^b + \beta h \sum_a S^a \right\}. \quad (6.12)$$

Contrary to the spherical case, where the single site average requires a free integration over continuous variables (giving rise to a $\mathrm{tr} \ln Q$ factor – see Chapter 5), here we have discrete $S^a = \pm 1$ variables, which involves slightly more labour.

To proceed with the computation we need at this point to make an ansatz on the structure of the overlap matrix Q^{ab}. As usual, we start looking at the most natural ansatz, the replica symmetric one.

[†] If instead of a free sum over i, j we keep to nearest neighbour interactions (nn) we need to introduce an overlap variable Q_i^{ab} which depends on the site. In this case, instead of (6.9), we get

$$\mathcal{L}\{Q_i^{ab}\} = -\frac{z(\beta J)^2}{2} \sum_{\substack{i,j \\ nn}} \sum_{(ab)} Q_i^{ab} (K^{-1})_{ij} Q_j^{ab} + \frac{Nn}{4}(\beta J)^2$$
$$+ \sum_i \ln \operatorname*{tr}_{\{S_i^a\}} \exp\left\{ (\beta J)^2 \sum_{(ab)} Q_i^{ab} S_i^a S_i^b + \beta h \sum_a S_i^a \right\},$$

where K_{ij} and its Fourier Transform in 1 Dimension are given by

$$K_{ij} = \begin{cases} 1 \text{ for } nn, \\ 0 \text{ otherwise}, \end{cases} \qquad \tilde{K}(k) = \sum_{\mu}^{2D} \cos k_\mu a,$$

a being the lattice spacing. The factor $z = 2D$ keeps track of the number of loops.

6.2 The RS ansatz

The RS ansatz is specified by setting

$$Q^{ab} = \left(1 - \delta_{a;b}\right) Q. \tag{6.13}$$

In this way we have

$$\sum_{ab} Q^{ab} S^a S^b = Q \left(\sum_a S^a \right)^2 - nQ \tag{6.14}$$

and therefore

$$\begin{aligned}
\zeta(S) &= \exp\left\{ \frac{(\beta J)^2}{2} Q \left(\sum_a S^a \right)^2 - \frac{n(\beta J)^2}{2} Q + \beta h \sum_a S^a \right\} \\
&= \int \frac{\mathrm{d}z}{\sqrt{2\pi}} \exp\left\{ -\frac{z^2}{2} + \beta J \sqrt{Q} z \sum_a S^a + \beta h \sum_a S^a - \frac{n(\beta J)^2}{2} Q \right\} \\
&= \int \frac{\mathrm{d}z}{\sqrt{2\pi}} \exp\left\{ -\frac{z^2}{2} - n \frac{(\beta J)^2}{2} Q \right\} \cosh^n\left(\beta J \sqrt{Q} z + \beta h\right),
\end{aligned} \tag{6.15}$$

where we have used a Hubbard–Stratonovich transformation to linearize the term quadratic in the spin variables. Thus, the saddle point equation (6.11) becomes

$$Q = \int \mathrm{D}z \tanh^2\left(\beta J \sqrt{Q} z + \beta h\right) \cosh^n\left(\beta J \sqrt{Q} z + \beta h\right), \tag{6.16}$$

where

$$\mathrm{D}z \equiv \frac{\mathrm{d}z}{\sqrt{2\pi}} \exp\left\{-z^2/2\right\}. \tag{6.17}$$

The free energy density and the local magnetization of the replicated system acquire the following expressions (for finite n):

$$m = \langle S^a \rangle = \int \mathrm{D}z \tanh\left(\beta J \sqrt{Q} z + \beta h\right) \cosh^n\left(\beta J \sqrt{Q} z + \beta h\right) \tag{6.18}$$

and

$$\begin{aligned}
n\beta f &= -\frac{\mathcal{L}}{N} = \frac{1}{4} (\beta J)^2 n(n-1) Q^2 - \frac{n}{4} (\beta J)^2 \\
&\quad - \ln \int \mathrm{D}z \exp\left\{ -\frac{n}{2} (\beta J)^2 Q \right\} \cosh^n\left(\beta J \sqrt{Q} z + \beta h\right).
\end{aligned} \tag{6.19}$$

The thermodynamic properties of the original system are finally recovered by performing the $n \to 0$ limit:

$$Q = \int \mathrm{D}z \tanh^2\left(\beta J \sqrt{Q} z + \beta h\right), \tag{6.20}$$

$$m = \int \mathrm{D}z \tanh\left(\beta J \sqrt{Q} z + \beta h\right), \tag{6.21}$$

$$\begin{aligned}\beta f = &-\frac{(\beta J)^2}{4} Q^2 + \frac{(\beta J)^2}{2} Q - \frac{(\beta J)^2}{4} \\ &- \int \mathrm{D}z \ln\left\{2\cosh\left(\beta J \sqrt{Q} z + \beta h\right)\right\}.\end{aligned} \tag{6.22}$$

If $h = 0$, $m = 0$ and Eq. (6.20) can be expanded for small Q (i.e. close to the transition line), giving

$$Q = (\beta J)^2 \, Q - \frac{2}{3} (\beta J)^4 \, Q^2 + \mathrm{O}(Q^4). \tag{6.23}$$

From this equation we can see that different solutions exist according to the sign of $[(\beta J)^2 - 1]$:

$$\begin{aligned} &1 - (\beta J)^2 > 0, \qquad Q = 0, \quad \textit{paramagnet}, \\ &1 - (\beta J)^2 < 0, \qquad Q \neq 0, \quad \textit{spin glass}, \end{aligned} \tag{6.24}$$

thus locating a transition at $T_c = J$.

However, one finds that below T_c the entropy becomes negative, showing that the replica symmetric ansatz does not provide a good mathematical description of the low temperature phase. Indeed, as we shall now discuss in detail, the RS saddle point is *not* stable below T_c and must therefore be discarded for stability reasons.

6.3 Stability around the RS saddle point

We recall that the replicated averaged partition function is given by

$$\overline{Z^n} = \int \prod_{(ab)} \left(\frac{\mathrm{d}Q^{ab}}{\sqrt{2\pi}}\right) \exp\{\mathcal{L}\{Q^{ab}\}\}. \tag{6.25}$$

The leading contribution to this integral in the thermodynamic limit is obtained by evaluating the effective Lagrangian $\mathcal{L}$ at a saddle point Q_{SP}, this is what we have done so far, by considering a replica symmetric saddle point whose structure in replica space is $Q^{ab} = \left(1 - \delta_{a;b}\right) Q$. If we want to investigate the stability of the saddle point we need to expand the Lagrangian around it and take into account the contribution of fluctuations. Thus we set

$$Q^{ab} = (Q^{ab})_{\mathrm{SP}} + \varphi_{ab}, \tag{6.26}$$

so that

$$\overline{Z^n} = \int \prod_{(ab)} \left(\frac{\mathrm{d}Q^{ab}}{\sqrt{2\pi}} \right) \exp \left\{ \mathcal{L}\{Q^{ab}_{\mathrm{SP}}\} - \frac{N}{2} \sum_{(ab)(cd)} \varphi_{ab} M^{ab;cd} \varphi_{cd} \right\}, \tag{6.27}$$

where we have introduced the Hessian matrix

$$M^{ab;cd} \equiv -\frac{1}{N} \frac{\partial^2 \mathcal{L}}{\partial Q^{ab} \partial Q^{cd}} |_{Q=Q_{\mathrm{SP}}}. \tag{6.28}$$

In particular, for the SK model and the RS saddle point studied so far, from Eq. (6.9) we find‡

$$\begin{aligned} \mathcal{L}\{Q^{ab}\} &= \mathcal{L}_{\mathrm{SP}}(Q) + \mathcal{L}_{\mathrm{fluct}}\{\varphi_{ab}\} \\ &= \mathcal{L}(Q) - N \frac{(\beta J)^2}{2} \left\{ \left(\frac{1}{(\beta J)^2} - 1 \right) \sum_{(ab)} \varphi^2_{ab} \right. \\ &\quad \left. - Q \sum_{(abc)} (\varphi_{ab}\varphi_{ac} + \varphi_{ab}\varphi_{bc}) - R \sum_{(ab) \neq (cd)} \varphi_{ab}\varphi_{bc} \right\}, \end{aligned} \tag{6.29}$$

where, as usual, (ab) indicates that the sum is over distinct pairs of indices, and (abc) over distinct triplets. The overlap Q is given by the saddle point equation (6.20), while the parameter R is given by

$$R = \int \mathrm{D}z \tanh^4 (\beta J \sqrt{Q} z + \beta h). \tag{6.30}$$

Note then that the fluctuation term in (6.29) has the general form

$$\begin{aligned} \mathcal{L}_{\mathrm{fluct}}\{\varphi_{ab}\} &\equiv -\frac{N}{2} \sum_{(ab)(cd)} \varphi_{ab} M^{ab;cd} \varphi_{cd} \\ &= -\frac{N}{2} \left\{ M_1 \sum_{(ab)} \varphi^2_{ab} + M_2 \sum_{(abc)} (\varphi_{ab}\varphi_{ac} + \varphi_{ab}\varphi_{bc}) + M_3 \sum_{(ab) \neq (cd)} \varphi_{ab}\varphi_{cd} \right\}, \end{aligned} \tag{6.31}$$

where $M^{ab;ab} \equiv M_1$, $M^{ab;ac} \equiv M_2$, and $M^{ab;cd} \equiv M_3$. In particular, for the SK model from Eq. (6.29), we have

$$\begin{cases} M_1 = 1 - (\beta J)^2, \\ M_2 = -(\beta J)^2 Q = -(\beta J)^2 \int \mathrm{D}z \tanh^2 (\beta J \sqrt{Q} z + \beta h), \\ M_3 = -(\beta J)^2 R = -(\beta J)^2 \int \mathrm{D}z \tanh^4 (\beta J \sqrt{Q} z + \beta h). \end{cases} \tag{6.32}$$

‡ When space is present, as in the previous footnote (in contradistinction to the SK model), the space dependence becomes explicit beyond mean field, at the fluctuation level, i.e. one has

$$Q^{ab}_i = Q^{ab}_{\mathrm{SP}} + \varphi^{ab}_i,$$

and the quadratic form of φ^{ab}_i has the same structure as (6.31) (see Section 8.2).

The stability of the saddle point can be examined by computing the eigenvalues of the Hessian matrix $M^{ab;cd}$. In other terms, this means re-writing the fluctuation term (6.31) in a diagonal form. The standard way to do that is the one followed by de Almeida and Thouless (1978). Here, however, we choose an alternative route and diagonalize the Hessian matrix by using the Replica Fourier Transform method that will be extensively used also later on, in the case of ansatzes with replica symmetry breaking.

6.3.1 The Replica Fourier Transform

As a starting point it is convenient to obtain the fluctuation term with unrestricted summations. This can be easily done if we exhibit the restricted summations as

$$\begin{aligned}
\sum_{(abc)} \varphi_{ab}\varphi_{ac} &= \frac{1}{2}\sum_{ab}\varphi_{ab}\sum_{c}\varphi_{ac}(1-\delta_{b;c}),\\
\sum_{(ab)\neq(cd)} \varphi_{ab}\varphi_{cd} &= \frac{1}{4}\sum_{ab}\sum_{cd}\varphi_{ab}\varphi_{cd}(1-\delta_{a;c})(1-\delta_{a;d})(1-\delta_{b;c})(1-\delta_{b;d})\\
&= \frac{1}{4}\sum_{ab}\sum_{cd}\varphi_{ab}\varphi_{cd}[1-(\delta_{a;c}+\delta_{a;b}+\delta_{b;c}+\delta_{b;d})+\delta_{a;c}\delta_{b;d}+\delta_{a;d}\delta_{b;c}],
\end{aligned} \tag{6.33}$$

where we also exploited the fact that $\varphi_{aa}=0$. In this way, we have

$$\begin{aligned}
\mathcal{L}_{\text{fluct}}\{\varphi\} = -\frac{N}{2}\Bigg\{&\frac{1}{2}(M_1-2M_2+M_3)\sum_{ab}\varphi_{ab}^2\\
&+2(M_2-M_3)\sum_{abc}\varphi_{ab}\varphi_{ac}+\frac{M_3}{4}\left(\sum_{ab}\varphi_{ab}\right)\left(\sum_{cd}\varphi_{cd}\right)\Bigg\}.
\end{aligned} \tag{6.34}$$

At this point we introduce a Fourier Transform over replica indices. We shall see that, just in the same way as the standard Fourier Transform directly diagonalizes fluctuation matrices with respect to space coordinates, the Replica Fourier Transform (RFT) does the same job in replica space.

For variables depending on a single replica index, the RFT and its inverse transformation are defined as

$$O_{\hat{a}} = \frac{1}{\sqrt{n}}\sum_{a} \mathrm{e}^{\frac{2\pi\mathrm{i}}{n}a\hat{a}} O_a; \qquad O_a = \frac{1}{\sqrt{n}}\sum_{\hat{a}} \mathrm{e}^{-\frac{2\pi\mathrm{i}}{n}a\hat{a}} O_{\hat{a}}, \tag{6.35}$$

with $a = 1, \ldots, n, \hat{a} = 0, 1, \ldots, n-1$ and a, $\hat{a}$ considered mod(n). For two replica variables, as the φ_{ab}, the generalization is straightforward and we have

$$\varphi_{ab} = \frac{1}{n} \sum_{\hat{a}\hat{b}} \mathrm{e}^{\frac{2\pi\mathrm{i}}{n}(a\hat{a}+b\hat{b})} \varphi_{\hat{a}\hat{b}},$$
$$\varphi_{\hat{a}\hat{b}} = \frac{1}{n} \sum_{ab} \mathrm{e}^{-\frac{2\pi\mathrm{i}}{n}(a\hat{a}+b\hat{b})} \varphi_{ab}. \tag{6.36}$$

Note the following relations (we use $\hat{0}$ when $\hat{a} = 0$):

$$\sum_a \mathrm{e}^{\frac{2\pi\mathrm{i}}{n} a\hat{a}} = n\,\delta_{\hat{a};\hat{0}}, \tag{6.37}$$

$$\sum_{ab} \varphi_{ab}^2 = \sum_{\hat{a}\hat{b}} \varphi_{\hat{a}\hat{b}} \varphi_{-\hat{a},\,-\hat{b}} = \sum_{\hat{a}\hat{b}} |\varphi_{\hat{a}\hat{b}}|^2, \tag{6.38}$$

$$\sum_{ab} \varphi_{ab} = n\,\varphi_{\hat{0}\hat{0}}. \tag{6.39}$$

Note also that, when transforming cubic (or quartic) terms the total replica 'momentum' is zero, exactly as for the ordinary space Fourier Transform.

If we Fourier Transform over replicas expression (6.31) we get

$$\mathcal{L}_{\text{fluct}} = -\frac{N}{2} \left\{ \frac{1}{2} M_{\mathrm{R}} \sum_{\hat{a}',\hat{b}'} |\varphi_{\hat{a}'\hat{b}'}|^2 + (M_{\mathrm{R}} + n\,(M_2 - M_3)) \sum_{\hat{a}'} |\varphi_{\hat{a}'\hat{0}}|^2 \right.$$
$$\left. + \frac{1}{2} \left(M_{\mathrm{R}} + 2n\,(M_2 - M_3) + \frac{n^2}{2} M_3 \right) \varphi_{\hat{0}\hat{0}}^2 \right\}, \tag{6.40}$$

where $M_{\mathrm{R}} \equiv M_1 - 2M_2 + M_3$, and primed variables assume only values different from zero (i.e. $\hat{a}' \neq \hat{0}$, $\hat{b}' \neq \hat{0}$).

This expression is almost in a diagonal form. However, the fields appearing in (6.40) are not yet orthogonal fields. The reason is that there is the constraint $\varphi_{aa} = 0$ that has not yet been taken into account. Let us see how we can handle this, by introducing new field variables. As a first step, by summing the constraint over all the replica indices, we get

$$\sum_a \varphi_{aa} = 0. \tag{6.41}$$

In terms of the RFT fields this equality becomes

$$\sum_{\hat{a}=0}^{n-1} \varphi_{\hat{a},-\hat{a}} = 0 = \varphi_{\hat{0}\hat{0}} + \sum_{\hat{a}'=1}^{n-1} \varphi_{\hat{a}',-\hat{a}'}, \tag{6.42}$$

from which we see that the sub-space of $\{\varphi_{\hat{a}',-\hat{a}'}\}$ is not orthogonal to $\varphi_{\hat{0}\hat{0}}$. To take care of that, let us introduce a set of new fields defined as

$$\Phi_{\hat{a}',-\hat{a}'} = \varphi_{\hat{a}',-\hat{a}'} + \frac{1}{n-1}\varphi_{\hat{0}\hat{0}}. \tag{6.43}$$

These are now orthogonal to $\varphi_{\hat{0}\hat{0}}$ and satisfy the constraint

$$\sum_{\hat{a}'=1}^{n-1} \Phi_{\hat{a}',-\hat{a}'} = 0. \tag{6.44}$$

Another relationship can be obtained if we RFT φ_{aa} over the index a and then sum over it:

$$\sum_a e^{\frac{2\pi i}{n} a\hat{\imath}'}\varphi_{aa} = 0, \tag{6.45}$$

where $\hat{\imath}' \neq 0$ (for $\hat{\imath} = 0$ we recover Eq. (6.41)). In terms of RFT fields this becomes

$$\sum_{\hat{a}} \varphi_{\hat{a},\hat{\imath}'-\hat{a}} = \varphi_{\hat{0}\hat{\imath}'} + \varphi_{\hat{\imath}'\hat{0}} + \sum_{\hat{a}''} \varphi_{\hat{a}'',\hat{\imath}'-\hat{a}''}, \tag{6.46}$$

where the double primed variable $\hat{a}''$ assumes values different from $\hat{0}$ and $\hat{\imath}'$. This relation shows that the sub-space of the $\{\varphi_{\hat{a}'',\hat{\imath}'-\hat{a}''}\}$ (with $\hat{\imath}'$ fixed) is not orthogonal to $\varphi_{\hat{0}\hat{\imath}'}$. We then proceed as before and define some new fields:

$$\Phi_{\hat{a}'',\hat{\imath}'-a''} = \varphi_{\hat{a}'',\hat{\imath}'-\hat{a}''} + \frac{1}{n-2}(\varphi_{\hat{0}\hat{\imath}'} + \varphi_{\hat{\imath}'\hat{0}}), \tag{6.47}$$

satisfying now

$$\sum_{\hat{a}''} \Phi_{\hat{a}'',\hat{\imath}'-\hat{a}''} = 0. \tag{6.48}$$

In terms of the new fields we can now write (De Dominicis and Pimentel, 1997, unpublished)

$$\begin{aligned}\mathcal{L}_{\text{fluct}} = -\frac{N}{2}\Bigg\{&\frac{1}{2}M_{\text{R}}\sum_{\hat{a}'=1}^{n-1}(\Phi_{\hat{a}',-\hat{a}'})^2 + M_{\text{L}}\frac{n}{2(n-1)}\varphi_{\hat{0}\hat{0}}^2 \\ &+ \sum_{\hat{\imath}'}\left(\frac{1}{2}M_{\text{R}}\sum_{\hat{a}''}|\Phi_{\hat{a}'',\imath'-\hat{a}''}|^2 + M_{\text{A}}\frac{n}{(n-2)}|\varphi_{\hat{0}\hat{\imath}'}|^2\right)\Bigg\},\end{aligned} \tag{6.49}$$

with

$$M_{\text{L}} = M_{\text{R}} + 2(n-1)(M_2 - M_3 + nM_3/4), \tag{6.50}$$

$$M_{\text{A}} = M_{\text{R}} + (n-2)(M_2 - M_3). \tag{6.51}$$

Now the fields are almost orthogonal (in the sense that they still have to obey the constraints (6.44) and (6.48)), and we can read directly from Eq. (6.49) the eigenvalues and their degeneracies. We can see that the fluctuation space is subdivided into three diagonal sub-spaces or 'sectors' which, following Bray and Moore (1978, 1979) we call: the *replicon* sector with eigenvalue M_{R}, the *longitudinal* sector with eigenvalue M_{L} and the *anomalous* sector with eigenvalue M_{A}. Their dimensions (i.e. the eigenvalue degeneracies) are given by

$$\begin{cases} \text{replicon:} & n\,(n-3)\,/2, \\ \text{anomalous:} & (n-1), \\ \text{longitudinal:} & 1. \end{cases} \tag{6.52}$$

In finite dimensions, when a field theory with space dependence appears, the fluctuation term that we have computed contributes to the Gaussian part of the effective space dependent Lagrangian. One has to take account of the space dependence of the fields, that can be diagonalized as usual with an ordinary Fourier Transform. In this way, e.g., $M_{\mathrm{R}} \to p^2 + M_{\mathrm{R}}$, where p^2 is the kinetic energy and M_{R} becomes the bare energy (or mass) of the replicon mode. From Eq. (6.49) we then see that, altogether, there are three masses, two of which are degenerate as $n \to 0$, given by

$$\begin{aligned} M_{\mathrm{R}} &= M_1 - 2M_2 + M_3, \\ M_{\mathrm{L}} &= M_1 - 4M_2 + 3M_3, \\ M_{\mathrm{A}} &= M_{\mathrm{L}}. \end{aligned} \tag{6.53}$$

The associated propagators G_{R}, G_{L} and G_{A} will be discussed in Chapter 9.

We can now immediately compute the three eigenvalues for the SK model. From Eq. (6.32) we find

$$M_{\mathrm{R}} = 1 - (\beta J)^2 \int \mathrm{D}z \left[1 - \tanh^2\!\left(\beta J \sqrt{Q}\, z + \beta h\right)\right]^2, \tag{6.54}$$

$$M_{\mathrm{L}} = M_{\mathrm{R}} + 2\,(\beta J)^2 \int \mathrm{D}z \tanh^2\!\left(\beta J \sqrt{Q}\, z + \beta h\right)\left[1 - \tanh^2\!\left(\beta J \sqrt{Q}\, z + \beta h\right)\right], \tag{6.55}$$

$$M_{\mathrm{A}} = M_{\mathrm{L}}. \tag{6.56}$$

From this we see that the masses for the longitudinal and anomalous sectors (degenerate as $n \to 0$) are always larger than zero. The replicon mass, on the other hand, is larger than zero in the paramagnetic region and vanishes on the AT transition line (de Almeida and Thouless, 1978) defined by $M_{\mathrm{R}} = 0$ (see Fig. 6.1).

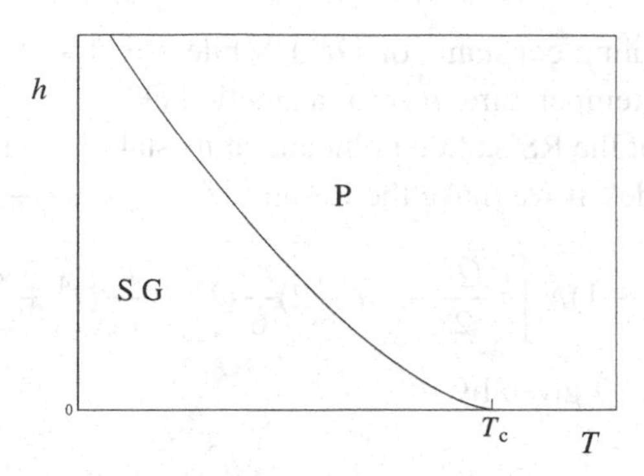

Figure 6.1 Phase diagram for the SK model: the continuous line is the AT line deduced by stability arguments

Below the AT line the RS solution gives a negative replicon eigenvalue and is therefore unstable.

6.3.2 Stability for the truncated model

So far we have considered as a starting point for our stability analysis the effective Lagrangian of the SK model as given in Eq. (6.9). An alternative route, particularly relevant for the finite dimensional treatment, is to look at the so-called *truncated* model (Bray and Moore, 1979). In this case one considers a small Q^{ab} expansion of $\mathcal{L}\{Q^{ab}\}$, i.e. the truncated Lagrangian

$$\begin{aligned}\mathcal{L}^{\mathrm{Tr}}\{Q^{ab}\} = N \Bigg\{ &-\frac{(\beta J)^2}{2}\sum_{(ab)}(Q^{ab})^2 + \frac{(\beta J)^4}{4}\mathrm{tr}\ Q^2 + \frac{(\beta J)^6}{6}\mathrm{tr}\ Q^3 \\ &+(\beta J)^8\left[\frac{1}{12}\sum_{ab}(Q^{ab})^4 + \frac{1}{8}\mathrm{tr}\ Q^4 - \frac{1}{4}\sum_{abc}(Q^{ab})^2(Q^{ac})^2\right] \\ &+\frac{(\beta J)^4}{2}h^2\sum_{ab}Q^{ab}\Bigg\},\end{aligned} \tag{6.57}$$

where usually only the first of the three quartic terms is retained (the crucial one to get the correct critical behaviour and the low temperature spin glass phase). The truncated Lagrangian then assumes the form (after an appropriate rescaling)

$$\mathcal{L}^{\mathrm{Tr}}\{Q^{ab}\} = N\left\{\frac{1}{2}\tau\ \mathrm{tr}\ Q^2 + \frac{w}{6}\mathrm{tr}\ Q^3 + \frac{u}{12}\sum_{ab}(Q^{ab})^4 + \frac{1}{2}h^2\sum_{ab}Q^{ab}\right\}, \tag{6.58}$$

where w, u are coupling constants of $O(1)$, while $\tau = 1 - T/T_c^0$, $T_c^0 = J$ being the (bare) transition temperature in zero magnetic field.

Let us now look at the RS saddle point and at its stability within the framework of the truncated model. If we make the RS ansatz $Q^{ab} = (1 - \delta_{a;b})Q$, we get

$$\mathcal{L}^{\mathrm{Tr}}(Q) = n(n-1)N\left[\tau\frac{Q^2}{2} + (n-2)\frac{w}{6}Q^3 + \frac{u}{12}Q^4 + \frac{h^2}{2}Q\right], \tag{6.59}$$

with the saddle point Q given by

$$\tau - wQ + \frac{u}{3}Q^2 + \frac{h^2}{2Q} = 0. \tag{6.60}$$

An expansion around the saddle point Q gives

$$\mathcal{L}^{\mathrm{Tr}}\{Q^{ab}\} = \mathcal{L}^{\mathrm{Tr}}(Q) - \frac{1}{2}N\left\{-2(\tau + uQ^2)\sum_{(ab)}\varphi_{ab}^2 - wQ\sum_{(abc)}(\varphi_{ab}\varphi_{ac} + \varphi_{ab}\varphi_{bc})\right\}. \tag{6.61}$$

This expression is of the same general form as Eq. (6.31), with

$$\begin{cases} M_1 = -2(\tau + uQ^2), \\ M_2 = -wQ, \\ M_3 = 0. \end{cases} \tag{6.62}$$

Using the general procedure described in the previous section, we easily get

$$M_{\mathrm{R}} = -2(\tau - wQ + uQ^2) = -\frac{4u}{3}Q^2 + \frac{h^2}{Q}, \tag{6.63}$$

$$M_{\mathrm{L}} = -\frac{4u}{3}Q^2 + 2wQ + \frac{2h^2}{Q}, \tag{6.64}$$

$$M_{\mathrm{A}} = M_{\mathrm{L}}. \tag{6.65}$$

As for the SK model, M_{L} and M_{A} are always positive, while the replicon mode vanishes on the AT line:

$$h^2 = \frac{4u}{3w^3}\tau^3 + \cdots, \tag{6.66}$$

where we have used that, from Eq. (6.63), $wQ = \tau + \mathrm{O}(\tau^2)$ at the transition. Note also that for $h = 0$, Q scales as τ/w close to T_c, so that there are two different mass scales:

(i) the replicon mode has a *small* mass $M_{\mathrm{R}} \sim \tau^2$;
(ii) the longitudinal and anomalous modes have a *large* mass scale M_{L}, $M_{\mathrm{A}} \sim \tau$.

6.4 Parisi Replica Symmetry Breaking

The problems of the RS solution, negative entropy and instability below the AT line induced people to look for different ansatzes.

The first step in a direction that was to lead to the correct answer was made by Blandin (1978; Blandin *et al.*, 1980), who proposed to break the replica symmetry by grouping replicas into blocks, with the order parameter Q^{ab} taking a value Q_1 in the blocks along the diagonal and a distinct value Q_0 in the off-diagonal blocks. As we shall see in the section below, this proposal was generalized by Parisi (1979, 1980) with blocks of size p_1 (instead of $p_1 = 2$ considered by Blandin) and with p_1 (together with Q_1 and Q_0) being kept as a variational parameter. Meantime, the interest was for a while focused on a solution proposed by Sommers (1978, 1979) which, at the price of introducing an anomaly in the linear response, was yielding a nonnegative entropy. His proposal was quickly translated into the above one-step RSB language (De Dominicis and Garel, 1979; Bray and Moore, 1980) with p_1 taken to infinity (and $Q_1 = Q_0$ with $\lim_{p_1\to\infty} p_1(Q_1 - Q_0) = -\Delta$, Δ being the anomaly). That solution was solving the problem with entropy, but was also shown to be unstable – a feature shared with the one-step Parisi ansatz and that can be removed by going to a multi-step RSB. A more serious and unwanted feature had to do with the fact that the corresponding overlap distribution $P(Q)$ (considered at the end of this section) was not positive, contrary to the $P(Q)$ derived within Parisi's ansatz.

6.4.1 The one-step ansatz

The analysis of the one-step RSB by Parisi (1979) was already very promising. The ansatz he proposed is like the one-step ansatz discussed in Chapter 5 for the p-spin model. The overlap matrix Q^{ab} is divided into n/p_1 blocks of size p_1 (p_1 is m in the notation of Section 5.2.5). Then matrix elements in the diagonal blocks are set to $Q^{ab} = Q_1$, while elements in the off-diagonal blocks are set to Q_0, i.e.

$$Q^{ab} = -Q_1\delta_{a;b} + \varepsilon_{a;b}(Q_1 - Q_0) + Q_0, \tag{6.67}$$

where $\varepsilon_{a;b}$ is one if the two replica indices a and b belong to the same block and zero otherwise, and, we recall, $Q^{aa} = 0$.

Then, the effective Lagrangian $\mathcal{L}$ of Eq. (6.9) (or, alternatively, (6.58) if we work with the truncated model) is extremized with respect to Q_1, Q_0 and p_1. In terms of energy and entropy this leads to

$$\frac{E}{N}(T = 0) = -0.765; \qquad \frac{S(T = 0)}{N} = -0.01; \tag{6.68}$$

to be compared on the one side with the numerical results $E/N = -0.763$; $S = 0$

and on the other side with the RS results $E/N = -0.798$; $S = -0.17$. One can see an obvious step in the right direction, with a less negative entropy.

From the point of view of the stability matrix, the (dangerous) replicon eigenvalue becomes smaller. For example, for the truncated model with the RS ansatz, as we have seen in the previous section, one has $M_\mathrm{R} \sim -4\tau^2/3$; this becomes $M_\mathrm{R} \sim -4\tau^2/27$ with a one-step breaking.

6.4.2 Multiple breaking

In a series of successive papers (Parisi, 1980a,b) Parisi then implemented a scheme of progressive breakings of the replica symmetry that eventually led to the correct solution. Note that two important requirements one has to impose when looking for RSB ansatzes are:

- replica equivalence, i.e. $\sum_b Q^{ab}$ must be independent of a;
- divisibility by n of the invariants occurring in the free energy, such as $\mathrm{tr}Q^2$, $\mathrm{tr}Q^3$, $\sum_{ab}(Q^{ab})^4$ etc., in order to have a finite contribution to the free energy in the limit $n \to 0$.

One easy way to satisfy these requirements is to have each row of Q^{ab} being a permutation of the other rows, which the one-step RSB just discussed obviously satisfies.

Let us now describe how this breaking can be generalized, taking into account these constraints. Starting from the one-step ansatz, we divide the diagonal blocks into p_1/p_2 sub-blocks, of size p_2. Then, in each diagonal block, we set $Q^{ab} = Q_2$ in the diagonal sub-blocks, and leave $Q^{ab} = Q_1$ in the off-diagonal sub-blocks. Starting now from the diagonal sub-blocks, this procedure can be iterated further, as shown in Fig. 6.2. Overall, if we consider an R-step RSB ansatz, this amounts to taking a sequence of R sizes

$$p_0 \equiv n > p_1 > p_2 > \cdots > p_R \geqslant p_{R+1} \equiv 1 \tag{6.69}$$

and take the matrix which has successive boxes with these sizes and with values $Q_0, Q_1, Q_2, \ldots, Q_R$ ($Q_{R+1} \equiv 0$), as displayed in Fig. 6.2.

An equivalent way to tell it is to draw a tree whose extremities are the n replicas, and with branches at level r with multiplicity p_r/p_{r+1}, as in Fig. 6.3.

In this figure we have also introduced a useful notation to exhibit how close are two replica indices in the tree: we say that $a \cap b = r$ if the 'leaves' corresponding to replica a and b have their closest common ancestor at the level r of the tree, r being zero at the root and R at the base. This is equivalent to saying that the overlap element $Q^{ab} = Q_r$ belongs to a block of size p_r, but *not* to a sub-block of size p_{r+1}. The higher r is, the closer the two replicas on the tree. The tree representation is useful because it displays the geometric properties of the overlap matrix, and in

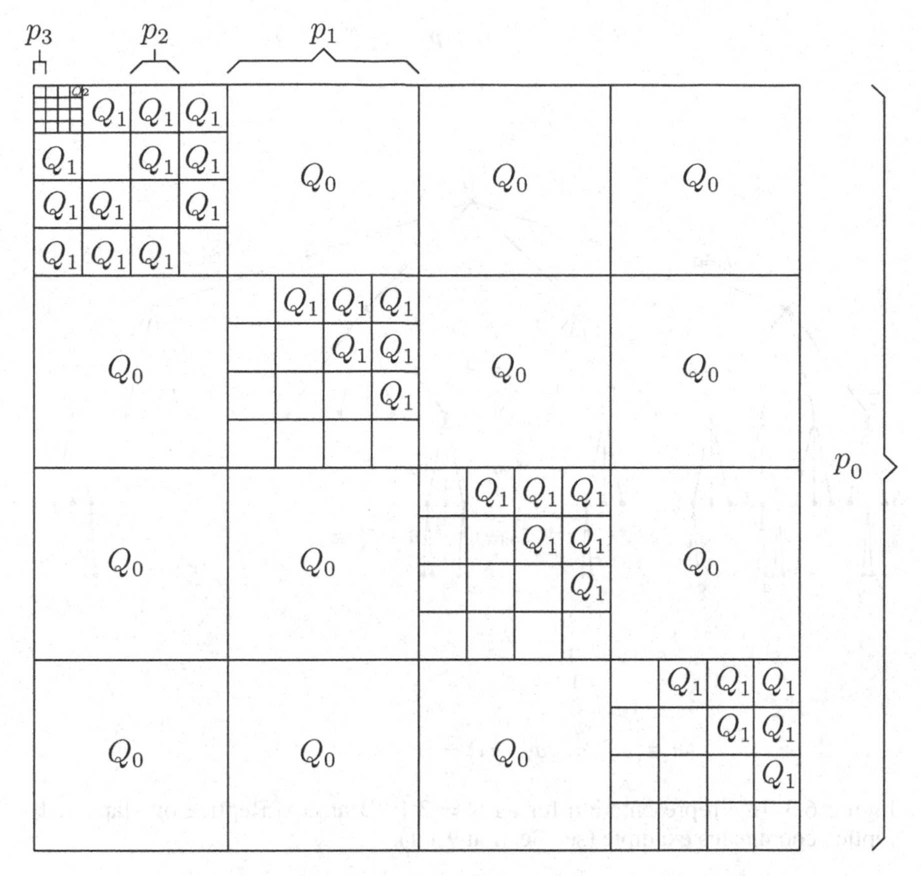

Figure 6.2 Overlap matrix with an $R = 3$ RSB ansatz

particular *ultrametricity*. That is, given three replicas abc then necessarily, out of Q^{ab}, Q^{bc}, Q^{ca}, at least two of them are equal (all triangles are isosceles, Q^{ab} being a co-distance).

Let us now proceed and compute the effective Lagrangian for the R-step RSB ansatz. As we shall see, the tree representation can tell us in a graphic way how to compute the invariants that occur in the free energy or derivatives thereof. For example, let us consider first the simplest quadratic term in the effective Lagrangian, namely $\operatorname{tr} Q^2 = \sum_a (\sum_b Q^{ab} Q^{ba})$. For an overlap $a \cap b = r$, the multiplicity of values, given a, that b can take is $p_r - p_{r+1}$, i.e. the size of the block p_r left over when block p_{r+1} is excluded. Thus we get

$$\begin{aligned}\operatorname{tr} Q^2 &= \sum_a \left(\sum_b Q^{ab} Q^{ba} \right) = n \sum_{r=0}^{R} (p_r - p_{r+1})\, Q_r^2 \\ &= n \left\{ (p_0 - p_1)\, Q_0 + (p_1 - p_2)\, Q_1 + \cdots + (p_R - p_{R+1})\, Q_R \right\}. \qquad (6.70)\end{aligned}$$

$$R = 3 \qquad n \equiv p_0 = \frac{p_0}{p_1}\,\frac{p_1}{p_2}\,\frac{p_2}{p_3}\,p_3$$

$$p_{R+1} = p_4 \equiv 1$$

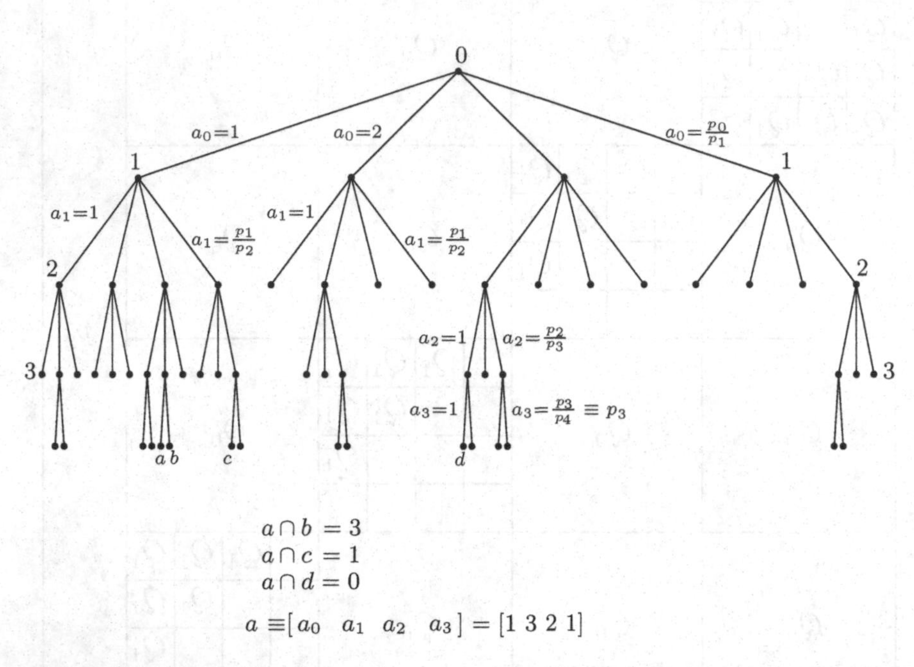

Figure 6.3 Tree representation for an $R = 3$ RSB ansatz. Replica overlaps and replica coordinates example (see Section 9.1.1).

Likewise, if we wish to compute $\sum_c Q^{ac}Q^{cb} \equiv [Q^2]_{ab}$ for a fixed overlap $a \cap b = r$, we have to consider three different geometries: (1) $c \cap a = c \cap b = t < r$; (2) $c \cap a = c \cap b = a \cap b = r$ for which the multiplicity is now $p_r - 2p_{r+1}$ since the overlap r is already occupied by a and b; (3) $c \cap a = r$, $c \cap b = t > r$ and $c \cap b = r$, $c \cap a = t > r$. These various cases are graphically shown in Fig. 6.4. Algebraically, we thus get for $a \cap b = r$

$$\sum_c Q^{ac}Q^{cb} = (Q^2)_r = \sum_{t=0}^{r-1} (p_t - p_{t+1})\,Q_t^2$$

$$+(p_r - 2p_{r+1})Q_r^2 + 2\,Q_r \sum_{t=r+1}^{R} (p_t - p_{t+1})Q_t. \qquad (6.71)$$

And with Eqs. (6.70) and (6.71) one immediately gets tr Q^3.

One can proceed in this way and express the effective Lagrangian $\mathcal{L}\{Q^{ab}\}$ and the free energy density in terms of the sets p_r, Q_r. Finally, from the saddle point

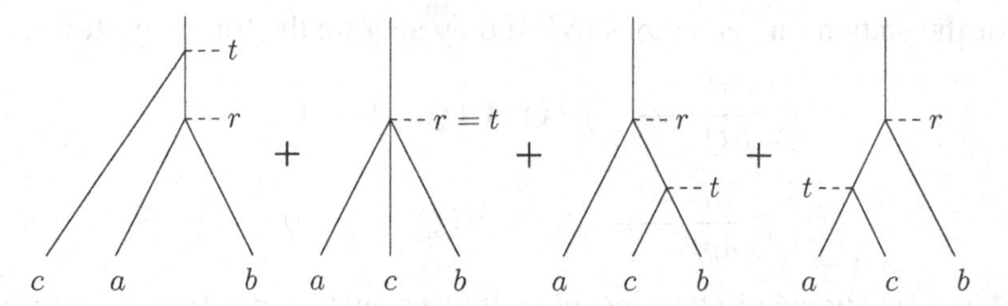

Figure 6.4 Various contributions to $\sum_c Q^{ac} Q^{cb}$. From left to right(see text): geometry (1), geometry (2) and the two cases of geometry (3).

equations we have

$$f = -\frac{1}{\beta} \lim_{n\to 0} \frac{1}{n} \left(\lim_{N\to\infty} \frac{1}{N} \mathcal{L}\{p_r, Q_r\} \right), \tag{6.72}$$

$$\frac{\partial \mathcal{L}}{\partial Q_r} = 0, \qquad \frac{\partial \mathcal{L}}{\partial p_r} = 0. \tag{6.73}$$

One gets a solution with

$$\begin{aligned} &Q_0 < Q_1 < Q_2 < \cdots < Q_R \qquad (Q_{R+1} \equiv 0), \\ &0 = n \equiv p_0 < p_1 < p_2 < \cdots < p_R < p_{R+1} \equiv 1. \end{aligned} \tag{6.74}$$

Note that, in this solution, the inequality (6.74) has reversed order and consequently the p_rs assume noninteger values. This is a consequence of the fact that we have considered the analytic continuation in the p_r for $n \to 0$ and it is an important feature to keep in mind. Indeed, it implies that the multiplicity $n(p_r - p_{r+1})$ associated with $Q_{ab} = Q_r$, that was positive with ordering (6.69), becomes now negative with (6.74). A negative multiplicity for the Qs entails a change in the sign of the convexity of the free energy $f(Q)$ at the stationary point. Likewise, a change in sign of the Hessian eigenvalue multiplicities will reverse the sign of the fluctuation contribution to the free energy.

Up to this point we have considered a finite number R of steps of replica symmetry breaking. The final move to get the exact solution is to take the limit to an infinite number of RSB, i.e. the limit $R \to \infty$, also called the 'full' RSB ansatz. This can be done easily starting from the finite R variational expression. If we call $x = r/(R+1)$, in the $R \to \infty$ limit we have

$$\begin{cases} p_r \to p(x), \\ Q_r \to Q(x), \\ p_r - p_{r+1} \to -\frac{\mathrm{d}p(x)}{\mathrm{d}x}\mathrm{d}x, \end{cases} \tag{6.75}$$

with x in the interval [0, 1], and where $p(x)$ is a continuous increasing function of x with $p(0) = 0$ and $p(1) = 1$.

As for the stationarity equations (6.73) they acquire the following form:

$$\frac{\partial f}{\partial Q_r} \to -\dot{p}(x)\left[\Phi\{Q, p\}\right] = 0, \tag{6.76}$$

$$\frac{\partial f}{\partial p_r} \to -\dot{Q}(x)\left[\Phi\{Q, p\}\right] = 0, \tag{6.77}$$

where Φ is a functional of $Q(x)$ and $p(x)$. It turns out that $p(x)$ can be chosen arbitrarily (but monotonically increasing and with $p(0) = 0$, $p(1) = 1$). The 'gauge' chosen by Parisi (1980) was the simplest one with $p(x) = x$.

6.4.3 Stationarity equations for the truncated model

Let us now focus upon the truncated model and write explicitly the RSB solution in the continuous limit (Parisi, 1980). From Eq. (6.58), by deriving with respect to Q^{ab}, we obtain the general stationarity equations

$$0 = 2\tau Q^{ab} + w \sum_c Q^{ac} Q^{cd} + \frac{2u}{3}(Q^{ab})^3 + h^2. \tag{6.78}$$

With an R-step ansatz and $a \cap b = r$, this becomes

$$\begin{aligned} 0 = 2\tau Q_r + w &\left\{ \sum_{t=0}^{r-1} (p_t - p_{t+1})\, Q_t^2 + (2p_r - p_{r+1})\, Q_r^2 \right. \\ &\left. + 2Q_r \sum_{t=r+1}^{R} (p_t - p_{t+1})\, Q_t \right\} + \frac{2u}{3} Q_r^3 + h^2, \end{aligned} \tag{6.79}$$

where we have used $Q_{R+1} \equiv 0$. In the $R \to \infty$ limit (in the Parisi gauge $p(x) = x$) one gets

$$0 = 2\tau Q(x) - w \left\{ \int_0^x \mathrm{d}t\; Q^2(t) + x\, Q^2(x) + 2Q(x) \int_x^1 \mathrm{d}t\; Q(t) \right\} + \frac{2u}{3} Q^3(x) + h^2. \tag{6.80}$$

Taking now a derivative with respect to x, we get

$$0 = 2\dot{Q}(x) \left\{ \tau - w \left[x Q(x) + \int_x^1 \mathrm{d}t\; Q(t) \right] + u\, Q^2(x) \right\}, \tag{6.81}$$

and finally, dividing by $\dot{Q}(x) \neq 0$ and taking once more a derivative,

$$0 = \dot{Q}(x) \{-wx + 2u\, Q(x)\}. \tag{6.82}$$

Hence, at least for those x for which $\dot{Q}(x) \neq 0$, we get

$$Q(x) = \frac{w}{2u} x. \tag{6.83}$$

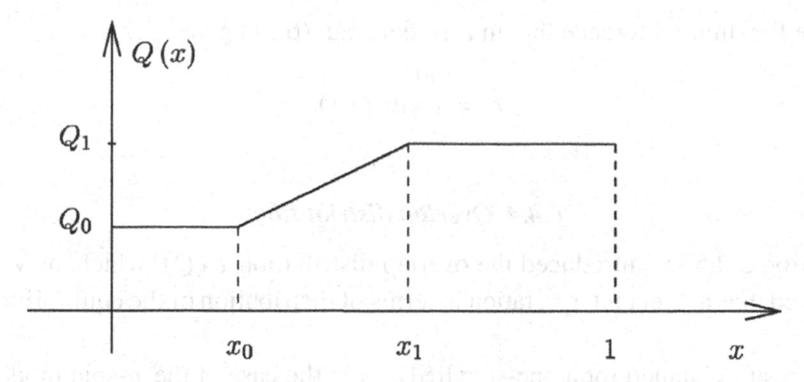

Figure 6.5 Overlap function $Q(x)$ for the truncated model

From Eq. (6.81) we see that the above behaviour cannot hold up to $x = 1$. Then Eq. (6.83) only holds up to a given $x = x_1$, while above

$$Q\,(x \geqslant x_1) \equiv Q_1 = \frac{w}{2u} x_1, \tag{6.84}$$

and from (6.81),

$$0 = \tau - w Q_1 + u Q_1^2. \tag{6.85}$$

Likewise, when $h^2 \neq 0$, from Eq. (6.80), we see that the linear behaviour cannot hold down to $x = 0$. Instead we have

$$Q\,(x < x_0) \equiv Q_0 = \frac{w}{2u} x_0, \tag{6.86}$$

where now Q_0 is fixed by Eq. (6.80),

$$0 = 2\tau Q_0 - 2w\, Q_0 \left[x_0 Q_0 + \frac{1}{2}\,(x_1 Q_1 - x_0 Q_0) + (1 - x_1)\, Q_1 \right] + \frac{2u}{3} Q_0^3 + h^2, \tag{6.87}$$

that is

$$0 = 2\tau Q_0 - \frac{4u}{3} Q_0^3 - 2Q_0 \left[w Q_1 - u Q_1^2 \right] + h^2 \tag{6.88}$$

and, with Eq. (6.85),

$$0 = -\frac{4u}{3} Q_0^3 + h^2, \tag{6.89}$$

thus yielding the complete $Q\,(x)$ behaviour (see Fig. 6.5). For $Q_0 = Q_1 \simeq \tau/w + \cdots$ this equation gives, as it should, the equation for the AT line $h^2 = (4u/3w^3)\tau^3 + \cdots$.

Note for future reference that in zero field Eq. (6.81) gives, for $x = 0$,

$$\tau = \int_0^1 \mathrm{d}t \; Q(t). \tag{6.90}$$

6.4.4 Overlap distribution

In Section 5.2.5 we introduced the overlap distribution $P(Q)$, which, as we have explained, has a direct interpretation in terms of distribution of the equilibrium pure states.

The result obtained for a one-step RSB, as in the case of the p-spin model, is

$$\overline{P_J(Q)} = \frac{n}{n(n-1)} \{(n - p_1)\,\delta(Q - Q_0) + (p_1 - 1)\,\delta(Q - Q_1)\} \tag{6.91}$$

(which is precisely Eq. (5.56) with $m \equiv p_1$). Note that given that the number of pairs $n(n-1)/2$ goes negative as $n \to 0$, it is consistent to accept the reversal of orderings implied by (6.74) that ensures thereby positiveness of the probability $P(Q)$.

The calculation can be immediately extended to the general case, yielding (Parisi, 1983),

$$\begin{aligned} P(Q) &= \frac{1}{(n-1)} \left\{ \sum_{t=0}^{R} (p_t - p_{t+1})\,\delta(Q - Q_t) \right\} \\ &= \int_0^1 \mathrm{d}x\; \delta(Q - Q(x)) = \int_{Q_0}^{Q_1} \mathrm{d}q\; \frac{\mathrm{d}x}{\mathrm{d}q}\; \delta(Q - q), \end{aligned} \tag{6.92}$$

that is

$$P(Q) = \frac{\mathrm{d}x(Q)}{\mathrm{d}Q}. \tag{6.93}$$

This equation is important. It gives a direct physical interpretation of the order parameter function $Q(x)$, or rather of the inverse function $x(Q)$. From the stationarity equations for $Q(x)$ we can explicitly compute the overlap distribution $P(Q)$, the result is shown in Fig. 6.6. We see that there is a finite probability of finding states with mutual overlap $Q \in [Q_0, Q_1]$, Q_0 representing the minimum overlap (maximum distance in phase space), and Q_1 the maximum one. Q_1 corresponds to the self-overlap of each equilibrium state. Thus, this model exhibits a highly nontrivial structure of states: given one of them, one finds other states at any value of the overlap between the self-overlap of the state itself and Q_0. Besides, we recall that the R-step RSB solution, and consequently also its $R \to \infty$ limit, has a remarkable geometric property: ultrametricity. In terms of states, this property implies

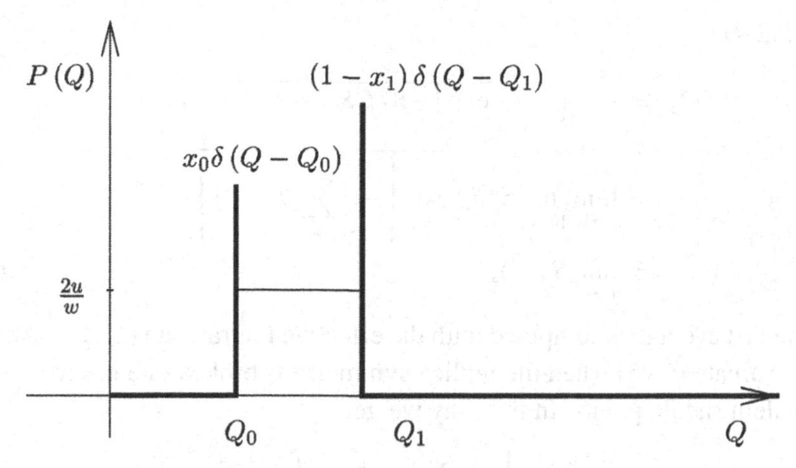

Figure 6.6 Overlap distribution function for the truncated model. The two vertical lines indicate two delta functions in Q_0 and Q_1, the heights being proportional to the weights x_0 and $1 - x_1$

a hierarchical structure which can be characterized in detail by considering more complex functions, such as the overlap probability distribution for triplets (Mézard *et al.*, 1984). A rough description of this structure is the following: for any value of the overlap $Q_0 < Q < Q_1$ the states are organized into disjoint clusters, such that any pair of states inside the same cluster has mutual overlap larger than Q. Then, one can divide each of these clusters by choosing a new overlap scale $Q' > Q$ and grouping together states with mutual overlap larger than Q'. This procedure can be iterated indefinitely. At any given scale Q, it can be shown that $x(Q)$ fully determines the weight distributions of clusters at that scale, much in the same way as what happens for states in the p-spin model where only one scale is present (see Section 5.2.5, where $m \equiv p_1$. Translated in the full-RSB notation p_1 is replaced by $p(x(Q_1))$, i.e. $x(Q_1)$ in the Parisi gauge).

6.4.5 Overlaps and susceptibilities

To conclude, let us look at the predicted behaviour of some measurable observables. Let us consider first the equilibrium spin overlap

$$Q_{\mathrm{eq}} = \frac{1}{N} \sum_i \overline{\langle S_i \rangle^2} = \overline{\langle S_i \rangle^2}, \tag{6.94}$$

where the second equality holds because disorder averages are site independent in the thermodynamic limit. The equilibrium spin overlap can be easily computed using the replica method, much in the same way we used for the $P(Q)$ (see

Section 5.2.4)

$$Q_{\text{eq}} = \frac{1}{Z^2} \operatorname*{tr}_{\{S_i\}} \operatorname*{tr}_{\{S'_i\}} \overline{\exp\{-\beta\mathcal{H}(S_i) - \beta\mathcal{H}(S'_i)\}(S_i S'_i)}$$

$$= \lim_{n\to 0} \operatorname*{tr}_{\{S_i^a\}} \overline{S_i^a S_i^b \exp\left\{-\beta \sum_a \mathcal{H}(S_i^a)\right\}}$$

$$= \lim_{n\to 0} \langle S^a S^b \rangle, \tag{6.95}$$

where the last average is computed with the effective Lagrangian (6.9). As we have seen, to evaluate (6.95) when the replica symmetry is broken one has to sum over all equivalent saddle points. In this way we get

$$Q_{\text{eq}} = \lim_{n\to 0} \frac{1}{n(n-1)} \sum_{ab} Q^{ab} = \int \mathrm{d}Q\, P(Q)\, Q. \tag{6.96}$$

Not surprisingly, Q_{eq} being defined in terms of the global equilibrium measure, it corresponds to the average value of the overlap between distinct pure states. On the other hand, ideally, an equilibrated system is going to be in one of the possible pure states. It is then convenient to define a more physical overlap for each individual state. This is the so-called Edwards–Anderson parameter

$$Q_{\text{EA}} = \frac{1}{N} \sum_i \overline{\langle S_i \rangle_\alpha^2}, \tag{6.97}$$

where α is a state index. Q_{EA} then corresponds to the self-overlap of an individual pure state and can reasonably be assumed to be self-averaging in the thermodynamic limit. It can also be computed with the replica method. Since the overlap distribution $P(Q)$ describes the distribution of overlaps between pure states, the self-overlap corresponds to the maximum possible value with finite probability, i.e.

$$Q_{\text{EA}} = Q(x_1). \tag{6.98}$$

The difference between Q_{eq} and Q_{EA} is a major consequence of the replica symmetry breaking.

Let us now consider susceptibilities. As for spin overlaps, different susceptibilities can be defined. If we consider the equilibrium local magnetization

$$m_i = \overline{\langle S_i \rangle} = \langle S^a \rangle \tag{6.99}$$

and differentiate with respect to an external magnetic field, we get the so-called *equilibrium* susceptibility

$$T\chi_{\text{eq}} = \sum_{ab} \langle S^a S^b \rangle = 1 + \frac{1}{n(n-1)} \sum_{a\neq b} Q^{ab}, \tag{6.100}$$

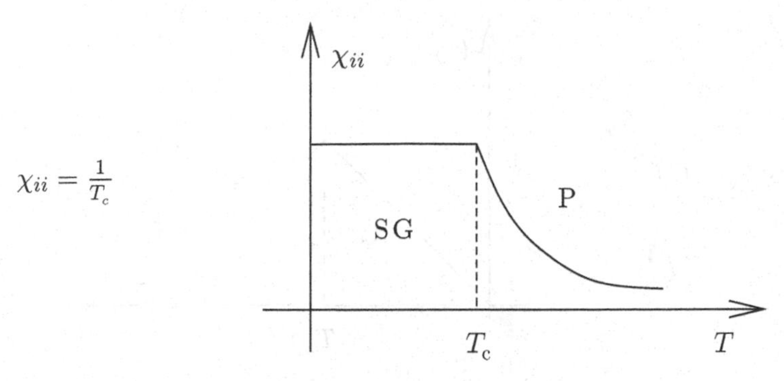

Figure 6.7 Equilibrium susceptibility

that becomes in the continuous limit

$$T\chi_{\text{eq}} = 1 - \int \mathrm{d}Q\, P(Q)\, Q = 1 - Q_{\text{eq}}$$
$$= 1 - \tau = \frac{T}{T_{\text{c}}}. \tag{6.101}$$

This susceptibility takes into account the contribution of all the states via their distribution $P(Q)$. Again, this is a consequence of using the global Boltzmann measure in a regime where ergodicity is broken. Its behaviour is shown in Fig. 6.7. From an experimental point of view, (6.101) is given to a very good approximation by the derivative of the so-called thermoremanent magnetization with respect to the magnetic field. Within this protocol, the system is cooled under a small magnetic field which is eventually switched off.

In contrast, the so-called EA susceptibility measures the susceptibility of each individual state and is defined as

$$T\chi_{\text{EA}} = \frac{1}{N}\sum_i \left(1 - \langle S_i \rangle_\alpha^2\right) = 1 - Q_{\text{EA}} = 1 - Q(x_1). \tag{6.102}$$

The last result can be directly understood. At $T = 0$, $Q(x_1) = 1$ and at low temperatures $Q = 1 - CT^2$, which gives approximately the behaviour shown in Fig. 6.8. This is like the linear response susceptibility, observed in experiments, which is given by the zero frequency limit of the a.c. susceptibility. In this case the system is quenched rapidly under zero magnetic field from high temperature to the working temperature below the transition, then a small oscillating field is applied that measures the susceptibility at that small frequency.

What is called the *SG susceptibility*,

$$\chi^{\text{SG}} = \frac{1}{N}\sum_i \overline{[\langle S_i S_j \rangle - \langle S_i \rangle \langle S_j \rangle]^2}, \tag{6.103}$$

will be considered in Chapter 8.

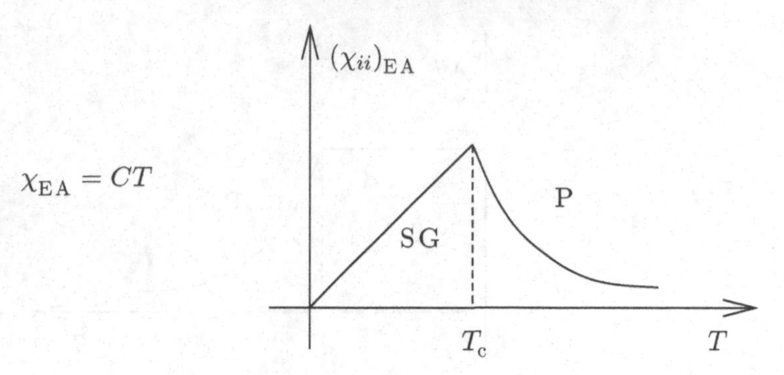

Figure 6.8 EA susceptibility

6.5 Summary

In this chapter we have introduced and discussed the Sherrington–Kirkpatrick model. For this model, the simplest RS ansatz does not yield the correct solution and one has to look for nontrivial patterns of replica symmetry breaking. The criteria necessarily satisfied for an acceptable solution are taken to be:

(i) the entropy must be positive,
(ii) the saddle point must be stable with respect to Gaussian fluctuations,
(iii) the overlap distribution must be positive.

The full RSB ansatz proposed by Parisi satisfies all these constraints and is widely accepted as the correct solution for the SK model. There have been in the past numerous numerical checks, and recently its validity has even been rigorously demonstrated (Guerra, 2003; Talagrand, 2006).

The physical scenario depicted by the Parisi RSB ansatz is of a novel kind. In the low temperature phase, ergodicity is broken in a nontrivial way and the phase space is organized into a hierarchical structure of pure states. This structure is described by the overlap distribution $P(Q)$, or, equivalently, by the function $Q(x)$ that represents the true order parameter for this model.

Note that, given the multiplicity of ergodic components at low temperature, thermodynamic averages performed with the Gibbs measure are not equivalent to averages inside one state but, rather, take into account the presence of all the states and of their weights. For instance, the equilibrium overlap $Q_{\rm eq} = (1/N) \sum_i \overline{\langle S_i \rangle^2}$ is given by $\int {\rm d}Q\, P(Q)\, Q$ and is *different* from the self-overlap $Q_{\rm EA}$ of each individual state. The same is true for response functions and susceptibilities, and one has to be careful in comparing theoretical values with results of experiments or numerical simulations.

References

Blandin A. (1978). *J. Phys. C*, **6**, 1499.
Blandin A, Gabay M. and Garel T. (1980). *J. Phys. C*, **13**, 403.
Bray A. J. and Moore M. A. (1978). *Phys. Rev. Lett.*, **41**, 1068.
Bray A. J. and Moore M. A. (1979). *J. Phys. C*, **12**, 79.
Bray A. J. and Moore M. A. (1980). *J. Phys. C*, **13**, L469.
De Almeida J. R. L. and Thouless D. J. (1978). *J. Phys. A.*, **11**, 129.
De Dominicis C. and Garel T. (1979). *J. Phys. Lett.*, **40**, L575.
Edwards S. F. and Anderson P. W. (1975). *J. Phys. F: Metal Phys.*, **5**, 965.
Guerra F. (2003). *Comm. Math. Phys.*, **233**, 1.
Mézard M., Parisi G., Sourlas N., Toulouse G. and Virasoro M. (1984). *J. Physique*, **45**, 843.
Mézard M., Parisi G. and Virasoro M. A. (1987). *Spin Glass Theory and Beyond*, Singapore, World Scientific, Vol. 9.
Parisi G. (1979). *Phys. Rev. Lett.*, **43**, 1754.
Parisi G. (1980a). *J. Phys. A*, **13**, L115; *J. Phys. A*, **13**, 1101.
Parisi G. (1980b). *J. Phys. A*, **13**, 1887.
Parisi G. (1983). *Phys. Rev. Lett.*, **50**, 1946.
Sherrington D. and Kirkpatrick S. (1975). *Phys. Rev. Lett.*, **35**, 1792.
Sommers H. J. (1978). *Z. Phys. B*, **31**, 301.
Sommers H. J. (1979). *Z. Phys. B*, **33**, 173.
Talagrand M. (2006). *Ann. Math.*, **163**, 221.

7
Mean field via TAP equations

In the previous chapter we have obtained the exact analytic solution of the SK model via the computation of the averaged static free energy using the replica method. The order parameter, the function $Q(x)$, encodes the nontrivial features of the thermodynamically relevant states at low temperature. In this chapter we focus onto another method, the so-called TAP approach (Thouless, Anderson and Palmer, 1977) which does not directly look at the averaged partition function. Rather, it is based on the formulation, for a given disorder realization, of mean field equations for the local magnetizations, in the same spirit of what is done for the Ising Model in the mean field approximation. In Section 7.1 we describe how the free energy functional can be derived, and what is its specific form in the case of the SK model. In Section 7.2 we come back to the concept of 'complexity', already encountered in Chapter 5, while in Section 7.3 we introduce the important property of supersymmetry. In Section 7.4, we demonstrate the equivalence between the TAP approach and the standard replica method for the thermodynamic quantities. Finally, in Section 7.5 we discuss the complexity of metastable states, the breaking of supersymmetry and its physical interpretation.

7.1 The TAP free energy and mean field equations

The TAP mean field free energy functional can be defined in the following way. Let us consider the system in the presence of an external magnetic field h_i:

$$\mathcal{H} = -\sum_{(ij)} J_{ij} S_i S_j - \sum_i h_i S_i . \tag{7.1}$$

Then, let us Legendre transform the associated (Helmholtz) free energy $-\beta F\{h_i\} = \ln Z\{h_i\}$, to obtain a function that only depends on the magnetization m_i.

$$\Gamma\{m_i\} = \min_{\{h_i\}} \left\{ F\{h_i\} + \sum_i h_i m_i \right\} , \tag{7.2}$$

where the local fields h_i fix the magnetizations at each site to their thermal expectation value $m_i = \langle S_i \rangle$ and obey the equation

$$\frac{\partial F\{h_i\}}{\partial h_i} = -m_i. \tag{7.3}$$

Once Γ is known, the value of the local magnetization is fixed by the inverse Legendre Transform relation

$$\frac{\partial \Gamma\{m_i\}}{\partial m_i} = h_i, \tag{7.4}$$

known as the TAP equations. One can introduce the so-called TAP free energy function $F_{\text{TAP}}\{m_i; h_i\} = \Gamma\{m_i\} - \sum_i h_i m_i$, whose minima obey Eq. (7.4) and therefore correspond to thermodynamically stable values of the local magnetization. Note that the TAP free energy is not a thermodynamic potential, but at the minima. Its physical meaning is that of a cost function, $\exp\{-\beta F_{\text{TAP}}\{m_i; h_i\}\}$, representing the probability of finding the local magnetizations with values m_i in the presence of external local magnetic fields h_i.

7.1.1 The TAP free energy for the SK model

There are various methods to compute the TAP free energy and TAP equations for the SK model (Thouless, Anderson and Palmer, 1977; De Dominicis, 1980; Plefka, 1982; Georges and Yedidia, 1991). Here we give just a short argument, based on the *cavity method* (Mézard, Parisi and Virasoro, 1986, 1987).

Let us consider a system with $N-1$ spins at equilibrium in a given state with local magnetization m'_j. Let us add a new spin at site i and assume that the new spin only slightly perturbs the existing state. Then, two things happen. The $N-1$ magnetizations m'_j generate a local field $\sum_j J_{ij} m'_j$ on site i that adds to the external field h_i, thus generating a magnetization

$$m_i = \tanh\left\{\beta \sum_j J_{ij} m'_j + \beta h_i\right\}. \tag{7.5}$$

On the other hand, the new site exerts back a field $J_{ij} m_i$ on each site j and, consequently, the local magnetization in j changes:

$$m_j = m'_j + \chi_{jj} J_{ji} m_i, \tag{7.6}$$

where the single-site susceptibility is given by

$$T\chi_{jj} = \langle S_j^2 \rangle - \langle S_j \rangle^2 = 1 - m_j^2. \tag{7.7}$$

Hence we get

$$m_i = \tanh\left\{\beta \sum_j J_{ij} m_j - \beta^2 \sum_j J_{ij}^2 m_i \left(1 - m_j^2\right) + \beta h_i\right\}$$
$$= \tanh\left\{\beta \sum_j J_{ij} m_j - \beta^2 m_i (1-q) + \beta h_i + \mathrm{O}\left(\frac{1}{N}\right)\right\}, \quad (7.8)$$

and the TAP equations (7.4) read

$$\beta h_i = \tanh^{-1} m_i - \beta \sum_j J_{ij} m_j - \beta^2 \sum_j J_{ij}^2 m_i \left(1 - m_j^2\right). \quad (7.9)$$

A more detailed analysis of higher order contributions (Mézard, Parisi and Virasoro, 1987; see also Bray and Moore, 1979; Plefka, 1982) shows that Eq. (7.9) must be supported by the stability condition

$$1 - \frac{\beta^2 J^2}{N} \sum_i \left(1 - m_i^2\right)^2 > 0, \quad (7.10)$$

which coincides with the De Almeida–Thouless stability condition $M_{\mathrm{R}} > 0$ found within the replica method, see Eq. (6.54). By functional integration over m_j one gets

$$\beta F_{\mathrm{TAP}}\{m_i; h_i\} = -\frac{1}{2}\beta \sum_{ij} J_{ij} m_i m_j - \frac{\beta^2}{4N} \sum_{ij} J_{ij}^2 \left(1 - m_i^2\right)\left(1 - m_j^2\right)$$
$$+ \frac{1}{2} \sum_i \left\{(1+m_i) \ln \frac{1}{2}(1+m_i) + (1-m_i) \ln \frac{1}{2}(1-m_i)\right\}$$
$$- \beta \sum_i m_i h_i, \quad (7.11)$$

provided one determines the integration constant (e.g. in the paramagnetic phase and zero field where $m_i = 0$).

7.1.2 The paramagnetic phase and the spin glass transition

Let us now analyze, at fixed disorder, the behaviour of the TAP equations. At high temperature, in the paramagnetic phase, the system of N coupled equations (7.4) only admits the solution $m_i = 0$. Lowering the temperature, we expect at some point the spin-glass transition to occur. This can be seen by looking at the Jacobian of Eq. (7.4), $\mathcal{J}_{ij} = \partial h_i / \partial m_j$, whose zero eigenvalues signal the occurrence of a transition.

For small m_i one has

$$\beta h_i \simeq m_i - \beta \sum_j J_{ij} m_j + \beta^2 m_i \sum_j J_{ij}^2 + O(m^2), \tag{7.12}$$

and the zero mode of $\mathcal{J}_{ij}$ can be exhibited by going to eigenmodes of J_{ij}:

$$\beta h_\lambda = m_\lambda [1 - \beta J_\lambda + \beta^2 J^2] + O(m^2), \tag{7.13}$$

where J_λ is an eigenvalue of J_{ij} and $J_{ij}^2 \sim J^2/N$. The first zero is hit for J_λ maximal: $J_\lambda^* = 2J$, and

$$\beta h_{\lambda^*} = m_{\lambda^*} [1 - \beta J]^2 + O(m^2),$$

signalling the transition at $T = J$.

For $T < T_c = J$, $m_i \neq 0$ but the global magnetization vanishes, $m = (1/N) \sum_i m_i = 0$. Also, it turns out that the TAP equations admit not one, but many solutions with a nontrivial free energy spectrum. If we associate minima of the TAP free energy with equilibrium (local or global) states, then the multiplicity of solutions in the TAP approach is related to the nontrivial structure of the thermodynamic order parameter that we have discussed in the last chapter. However, as we shall see in the next sections, the TAP approach allows us to investigate also metastable states, which are not captured by a purely thermodynamical method.

7.2 Number of solutions and complexity

The TAP equations may admit many solutions. In particular, for the spin glass models studied so far, the number of solutions is exponentially high in the size of the system, i.e. $\mathcal{N} \sim \exp(N\Sigma)$, where Σ is an entropy and is usually called *complexity*. A more refined quantity is the number density, that is the number of TAP solutions with free energy density f belonging to $[f, f + \mathrm{d}f]$. Let us denote with α a particular solution m_i^α, with free energy $F_{\mathrm{TAP}}\{m_i^\alpha\}$ (we drop the dependence on $\{h_i\}$ from now on). Then we have (for any given sample)

$$\mathcal{N}(f) = \sum_{\alpha=1}^{\mathcal{N}} \delta[F_{\mathrm{TAP}}\{m_i^\alpha\} - Nf]. \tag{7.14}$$

In general also the number density is exponential in the size of the system, and the complexity $\Sigma(f)$ acquires a positive value in a well defined (temperature dependent) free energy density interval. To explicitly estimate the complexity one has to perform averages over the disorder distribution. The simplest thing to do is to average directly the number density to get the so-called ‘annealed’ complexity $\Sigma_{\mathrm{an}}(f) = (1/N) \ln \overline{\mathcal{N}(f)}$. However, the number density, not being an extensive quantity, fluctuates much from sample to sample and consequently Σ_{an} corresponds to the *typical* complexity only at high energies (see the discussion on the REM model in Chapter 5). The correct estimate is in general given by the 'quenched' computation $\Sigma(f) = (1/N)\overline{\ln \mathcal{N}(f)}$, which is, however, more difficult to perform, involving a logarithm and thus requiring the use of replicas.

For the time being, let us skip the technical details related to the computation of averages and rather focus on the physical interpretation. Usually for spin glass systems the complexity $\Sigma(f)$ is finite for $f \in [f_0, f_m]$ (see the REM and the p spin cases previously analyzed), being zero at the lower band edge f_0 and displaying a maximum at a given free energy density $f_{\rm m}$. If we can check that the TAP solutions counted are actually the stable ones (i.e. minima of $F_{\rm TAP}$), then we can conclude that the model has a nontrivial spectrum of metastable states (i.e. locally stable states with free energy density higher than the ground state, f_0 in the above notation). If we are interested in the thermodynamic properties, we must in principle take into account all these states and their relative weights. In terms of TAP states, the global partition function of the system is given by

$$Z_{\rm TAP} = \sum_{\alpha}^{\mathcal{N}} \exp\left\{-\beta F_{\rm TAP}\{m_i^\alpha\}\right\} = \int {\rm d}f \, \exp[-\beta N(f - T\Sigma(f))]. \quad (7.15)$$

In the thermodynamic limit, the integral is dominated by the maximum f^* of the exponent and we find for the free energy of the system

$$\frac{F}{N} = f^* - T\Sigma(f^*),$$
$$\beta = \left.\frac{\partial \Sigma(f)}{\partial f}\right|_{f^*} \quad (7.16)$$

(where, of course, one has to use the quenched complexity to evaluate f^* and F). For the SK model it turns out that the partition function (7.15) is always dominated by the ground states, i.e. $f^* = f_0$, indicating that the metastable states are thermodynamically irrelevant. In other terms, if we associate with each TAP state a weight

$$W_\alpha = \exp\left[-\beta F_{\rm TAP}\{m_i^\alpha\}\right] / Z_{\rm TAP}, \quad (7.17)$$

only the ground states have finite weight in the thermodynamic limit. These states have zero complexity, i.e. $\Sigma(f^*) = 0$. Thus, even if numerous, they can be at most algebraically increasing with N and are anyway entropically sub-dominant with respect to all the others. Given the weights (7.17) one is in principle able to re-build the whole thermodynamics, the average value of a generic observable O being computed as

$$\langle O \rangle_{\rm TAP} = \sum_\alpha W_\alpha O\{m_i^\alpha\}. \quad (7.18)$$

Of course, we expect the TAP averages (7.18) to be consistent with the averages computed directly with the Boltzmann measure, and the TAP approach to give the same results as the replica method (as applied, for example, in the previous chapter). More specifically, the free energy density of the system must be the same, $(1/N)\ln Z_{\text{TAP}} = (1/N)\ln Z$, that is

$$\frac{1}{N}\ln\left(\sum_{\{S_i\}}\exp[-\beta\mathcal{H}\{S_i\}]\right) = \frac{1}{N}\ln\left(\sum_{\alpha=1}^{\mathcal{N}}\exp\left[-\beta F_{\text{TAP}}\{m_i^\alpha\}\right]\right). \quad (7.19)$$

This consistency is not trivial to show. For the SK model it was first proved by De Dominicis and Young in 1983. A more general proof, based on the BRST supersymmetry, will be sketched in the next sections.

The TAP approach gives access also to metastable states that are not thermodynamically relevant. One may wonder whether these states play a role for the dynamics, as happens for the p-spin model where the asymptotic dynamics is asymptotically confined at the level of the threshold states (the states with highest complexity). It turns out that for the SK model this is not the case, the dynamics always approaches the ground states, even in the off-equilibrium regime, indicating that metastable TAP states are not dynamically trapping. The reason for that, as we shall see in the next sections, lies in a lack of stability and an extreme sensitivity to external perturbations.

7.3 The BRST supersymmetry

The number density can be written also in integral form as

$$\begin{aligned}\mathcal{N}(f) &= \sum_{\alpha=1}^{\mathcal{N}}\int\prod_i\left(\mathrm{d}m_i\ \delta\left(m_i - m_i^\alpha\right)\right)\ \delta[F_{\text{TAP}}\{m_i\} - Nf] \\ &= \int\prod_i\left(\mathrm{d}m_i\ \delta(\partial_i F_{\text{TAP}}\{m_i\})\right)\ |\det(\partial_i\partial_j F_{\text{TAP}}\{m_i\})\ |\ \delta[F_{\text{TAP}}\{m_i\} - Nf],\end{aligned} \quad (7.20)$$

where the delta functions implement the TAP equations $\partial_i F_{\text{TAP}}\{m_i\} = 0$, and the determinant of the Jacobian is needed as normalization. This expression is the starting point for the analytic computation of the averaged complexity, both the annealed and the quenched ones. In (7.20), however, the modulus of the determinant is quite hard to handle, from a mathematical point of view. Most of the calculations of the complexity assume that the modulus can in fact be disregarded and this is what we shall do in the following. We note that this approximation is a priori a dangerous one, since without the modulus each TAP solution is weighted with the sign of the Hessian determinant, with the risk of uncontrolled cancellations. Indeed, if we do not impose any constraint on the free energy density and count all the TAP solutions, dropping the modulus in Eq. (7.20) gives a topological constant by virtue of the Morse theorem, instead of the correct result $\mathcal{N} \sim \exp[N\Sigma]$. However, if we fix an energy density level we may hope that only a given class of TAP solutions (e.g. minima, or saddles of a given order) dominate at that level. In particular, if the free energy density is low enough, we expect only minima of F_{TAP} to dominate. In this case, when the Hessian has a well defined sign, the modulus becomes redundant and can be dropped. In any case, the previous argument must be handled with care, and always checked a posteriori. For the p-spin model, for example, the hypothesis of dominance of TAP solutions is well under control (Cavagna, Garrahan and Giardina, 1998), while, as we shall see, for the SK model the scenario is much more subtle. Let us now proceed and further develop Eq. (7.20). We can use an exponential representation for the delta functions and the determinant,

$$\prod_i \delta(\partial_i F_{\text{TAP}}) = \int_{-i\infty}^{+i\infty} \prod_i (\mathrm{D}x_i) \exp\left(\sum_i x_i \partial_i F_{\text{TAP}}\{m_i\}\right), \tag{7.21}$$

$$\delta(F_{\text{TAP}} - f) = \int_{-i\infty}^{+i\infty} \mathrm{D}u \, \exp[u\,(F_{\text{TAP}}\{m_i\} - Nf)], \tag{7.22}$$

$$\det(\partial_i \partial_j F_{\text{TAP}}) = \int_{-\infty}^{+\infty} \prod_i (\mathrm{D}\bar{\psi}_i \, \mathrm{D}\psi_i) \exp\left[\sum_{ij} \bar{\psi}_i \psi_j \partial_i \partial_j F_{\text{TAP}}\{m_i\}\right], \tag{7.23}$$

where $\{\bar{\psi}_i, \psi_i\}$ are anti-commuting Grassmann variables, and $\mathrm{D}x_i$, $\mathrm{D}u$ etc. stand for $\mathrm{d}x_i/\sqrt{2\pi}$, $\mathrm{d}u/\sqrt{2\pi}$ etc. In this way we can write (Zinn-Justin, 1989),

$$\mathcal{N}(f) = \int \mathrm{D}u \, \mathrm{e}^{-\beta u N f} \int \prod_i (\mathrm{D}m_i \, \mathrm{D}x_i \, \mathrm{D}\bar{\psi}_i \, \mathrm{D}\psi_i) \mathrm{e}^{\mathcal{L}\{m_i, x_i, \bar{\psi}_i, \psi_i\}} \tag{7.24}$$

with the Lagrangian $\mathcal{L}$ given by,

$$\mathcal{L}\{m_i, x_i, \bar{\psi}_i, \psi_i\} = \sum_i x_i \partial_i F_{\text{TAP}}\{m_i\} + \sum_{ij} \bar{\psi}_i \psi_j \partial_i \partial_j F_{\text{TAP}}\{m_i\} + u F_{\text{TAP}}\{m_i\}. \tag{7.25}$$

A key property of the Lagrangian (7.25) is its invariance under a generalization of the Becchi–Rouet–Stora–Tyutin (BRST) supersymmetry (Becchi *et al.*, 1975; Tyutin, 1975; Kurchan, 1991). If ϵ is an infinitesimal Grassmann parameter, it is straightforward to verify that (7.25) is invariant under the following transformation:

$$\delta m_i = \epsilon\, \psi_i, \qquad \delta x_i = -\epsilon\, u\, \psi_i, \qquad \delta \bar{\psi}_i = -\epsilon\, x_i, \qquad \delta \psi_i = 0$$
$$\Rightarrow \qquad \delta \mathcal{L} = 0. \tag{7.26}$$

The BRST invariance has interesting consequences on average values. The fact that $\delta\mathcal{L} = 0$ under the BRST supersymmetry implies that the average of any observable of the same variables performed with this Lagrangian must be invariant too. For an observable $\mathcal{O}$,

$$\langle \mathcal{O}(\Lambda) \rangle = \int \mathrm{D}\Lambda\, \mathcal{O}(\Lambda)\, \mathrm{e}^{S(\Lambda)}, \tag{7.27}$$

with $\Lambda = \{m_i, x_i, \bar{\psi}_i, \psi_i\}$, we have

$$\langle \mathcal{O}(\Lambda) \rangle - \langle \mathcal{O}(\Lambda - \delta\Lambda) \rangle = \langle \delta\mathcal{O}(\Lambda) \rangle = 0. \tag{7.28}$$

This property can be used to generate some useful Ward–Takahashi identities. Fruitful choices are $\mathcal{O} = m_i^k \bar{\psi}_j$ and $\mathcal{O} = x_i^k \bar{\psi}_j$, whose variation gives

$$k \left\langle m_i^{k-1} \bar{\psi}_i \psi_j \right\rangle = -\left\langle m_i^k x_j \right\rangle, \tag{7.29}$$

$$k \left\langle x_i^{k-1} w' \bar{\psi}_i \psi_j \right\rangle = \left\langle x_i^k x_j \right\rangle. \tag{7.30}$$

If one starts from a replicated version of (7.24), which is needed to compute the quenched complexity, the BRST supersymmetry can be easily shown to hold for the corresponding replicated Lagrangian and the Ward–Takahashi identities (7.29), (7.30) hold for fields bearing replica indices (Annibale *et al.*, 2003a,b). These BRST identities do play a crucial role in the computation of the complexity and of TAP averages. They encode some conditions that must be implemented to ensure the consistency of thermodynamics calculations. More than that, they not only are mathematical relations between abstract parameters, but also have a deep physical meaning which is crucial to understanding the structure of metastable states.

7.4 Thermodynamic properties and equivalence to the replica method

Let us now discuss in more details the consistency of the TAP approach with standard thermodynamics. As we have seen, this means proving that the global free energy of the system is the same if computed with the TAP weights and with the Boltzmann measure, see Eq. (7.19). When averaging over the quenched disorder this implies an equivalence between the replicated partition functions

$$\overline{Z^n(J_{ij})} = \overline{Z^n_{\mathrm{TAP}}(J_{ij})}. \tag{7.31}$$

This identity can be demonstrated by introducing replicas and overlap matrices in both sides much in the same way as was done in the last chapter. For the left hand side we find the replicated effective Lagrangian (6.9), expressed in terms of the spin overlap matrix $Q^{ab} = \langle S^a S^b \rangle$. For the right hand side, starting from the integral representation (7.24), we also find an effective Lagrangian that can be expressed in terms of an overlap matrix, this time involving magnetizations instead of configurations, $M^{ab} = \langle m^a m^b \rangle$. It is possible to show that these two effective Lagrangians are formally identical, provided that some necessary conditions between the variational parameters entering the Lagrangians are satisfied (De Dominicis and Young, 1983; Cavagna *et al.*, 2003). Since these conditions have a clear physical interpretation, it is interesting to look at them explicitly.

From the very definition of equilibrium magnetization and energy we trivially have

$$\frac{1}{\beta}\frac{\partial Z}{\partial h_i} = \sum_{\{S_i\}} S_i \exp[-\beta\mathcal{H}\{S_i\}] = \langle S_i\rangle\, Z, \tag{7.32}$$

$$-\frac{\partial Z}{\partial \beta} = \sum_{\{S_i\}} \mathcal{H}\{S_i\} \exp[-\beta\mathcal{H}\{S_i\}] = \langle \mathcal{H}\rangle\, Z. \tag{7.33}$$

On the other hand, from (7.19) we get,

$$\begin{aligned}\frac{1}{\beta}\frac{\partial Z_{\mathrm{TAP}}}{\partial h_i} &= X_i - \sum_{\alpha}^{\mathcal{N}(h,\beta)} \frac{\partial F_{\mathrm{TAP}}\{m_i^\alpha\}}{\partial h_i} \exp\left[-\beta F_{\mathrm{TAP}}\{m_i^\alpha\}\right]\\ &= X_i + \langle m_i\rangle_{\mathrm{TAP}}\, Z_{\mathrm{TAP}},\end{aligned} \tag{7.34}$$

$$\begin{aligned}-\frac{\partial Z_{\mathrm{TAP}}}{\partial \beta} &= Y + \sum_{\alpha}^{\mathcal{N}(h,\beta)} \frac{\partial F_{\mathrm{TAP}}\{m_i^\alpha\}}{\partial \beta} \exp\left[-\beta F_{\mathrm{TAP}}\{m_i^\alpha\}\right]\\ &= Y + \langle E\rangle_{\mathrm{TAP}}\, Z_{\mathrm{TAP}},\end{aligned} \tag{7.35}$$

where X_i and Y are the contributions coming, respectively, from the dependence on h and β of the total number of TAP states, $\mathcal{N}(h, \beta)$. From the above equations we can see that a minimal requirement for the two approaches to be consistent is that these extra factors X_i and Y do vanish, and indeed $X_i = Y = 0$ is precisely the condition necessary to demonstrate the equivalence of the replicated Lagrangians coming from (7.31). This condition is trivially satisfied when the dominant TAP states have zero complexity, as in the SK model (and indeed this is what De Dominicis and Young noted in their paper). However, a more general and elegant argument can be done using supersymmetry, which is also valid when the dominant states are not the ground states, as for example in the p-spin. With a little algebra one can indeed obtain (Cavagna *et al.*, 2003)

$$\begin{aligned}X_i &= \langle x_i\rangle,\\ Y &= \left\langle \sum_i \left[x_i + \bar{\psi}_i\psi_i\frac{\partial}{\partial m_i}\right]\frac{\partial}{\partial\beta}\frac{\partial F_{\mathrm{TAP}}}{\partial m_i}\right\rangle\\ &= \sum_i\sum_k c_k^i\langle m_i^k x_i\rangle + k\,\langle m_i^{k-1}\bar{\psi}_i\psi_i\rangle,\end{aligned} \tag{7.36}$$

with

$$c_k^i = \frac{1}{k!} \frac{\partial^k}{\partial^k m_i} \frac{\partial}{\partial \beta} \frac{\partial F_{\text{TAP}}}{\partial m_i}. \tag{7.37}$$

A comparison between (7.36) and (7.29) then clearly shows that the relations $X_i = 0$ and $Y = 0$ are a direct consequence of the BRST symmetry.

The equivalence (7.31) between the TAP approach and the standard thermodynamics can then be fully demonstrated. As a consequence, the interpretation of TAP minima as thermodynamic states is well grounded, at least for what concerns the thermodynamically dominant TAP minima with free energy density f^*. Equation (7.31) also yields a direct relationship between the self-overlap of the dominant states $M_{aa} = Q_{\text{EA}} = (1/N) \sum_i \langle (m_i^\alpha)^2 \rangle_{\text{TAP}}$ and the maximal value of Q^{ab}, i.e. $Q_{\text{EA}} = Q(x_1)$.

7.5 Metastable states and supersymmetry breaking

Let us now discuss the other TAP states, i.e. minima of the mean field free energy with $f \neq f^*$. Information on these states can be obtained via the computation of the complexity $\Sigma(f)$ starting from Eq. (7.24). For models like the p-spin this computation is straightforward and the results have been already discussed in Chapter 5. For the SK model, on the other hand, the analysis is more subtle. In particular, it turns out that the correct computation of $\Sigma(f)$ (Bray and Moore, 1980) involves a solution (at the saddle point) which is *not* supersymmetric (Cavagna *et al.*, 2003; see also Crisanti *et al.*, 2003). In other terms, the supersymmetry that we have discussed in Section 7.3 and that does hold for the thermodynamic states with $f = f^* = f_0$ is spontaneously broken at higher free energy density. This was somewhat of a surprise, but has been confirmed both with analytical arguments (Aspelmeier, Bray and Moore, 2004; Crisanti *et al.*, 2004) and with numerical simulations (Cavagna, Giardina and Parisi, 2003). The surprise comes from the fact that the supersymmetry not only is a formal property of the calculation, but also has a precise physical interpretation. This interpretation, as we shall see, is also the key to understanding what happens when there is a spontaneous breaking.

7.5.1 BRST supersymmetry and fluctuation–dissipation relations

Let us consider again the Ward–Takahashi Identities generated by the supersymmetry. In particular, let us focus on the Eq. (7.29) and set $k = 1$, then we get

$$\langle m_i x_j \rangle + \langle \bar{\psi}_i \psi_j \rangle = 0. \tag{7.38}$$

From the Lagrangian as in (7.25) one can write (see e.g. Parisi, 2006),

$$\langle \bar{\psi}_i \psi_j \rangle = \left\langle \left[\partial\partial F_{\mathrm{TAP}}\{m_i\} \right]_{ij}^{-1} \right\rangle, \tag{7.39}$$

identifying the second term of (7.38) as one component of the inverse Hessian. As for the first term of (7.38), it suffices to take the derivative of the average local magnetization with respect to the external magnetic field h_j, to obtain

$$\left. \frac{\mathrm{d}\langle m_i \rangle}{\mathrm{d}h_j} \right|_{h=0} = -\langle m_i x_j \rangle. \tag{7.40}$$

Here all averages are performed with the measure appearing in Eq. (7.24), that is averages over all the TAP states with free energy density f.

Thus it becomes clear that the Ward–Takahashi Identity (7.38) gives the relationship between the susceptibility and the curvature of the minima of the TAP free energy,

$$\left. \frac{\mathrm{d}\langle m_i \rangle}{\mathrm{d}h_j} \right|_{h=0} = \left\langle \left[\partial\partial F_{\mathrm{TAP}}\{m_i\} \right]_{ij}^{-1} \right\rangle, \tag{7.41}$$

which is nothing but the *static Fluctuation–Dissipation Theorem.*

7.5.2 Supersymmetry violation and fragility of the structure of states

Thus, a spontaneous breaking of the supersymmetry implies a violation of the static fluctuation–dissipation theorem. The reason for that lies in the peculiar structure of metastable states for the SK model. Recent studies (Aspelmeier, Bray and Moore, 2004; Cavagna, Giardina and Parisi, 2004; Parisi and Rizzo, 2004) have shown that at low temperatures *all* the stationary points of $F_{\mathrm{TAP}}\{m_i\}$ are organized into minimum–saddle pairs. The minimum and the saddle are connected along a mode that is the softer the larger the system size N. Moreover, the free energy difference of the paired stationary points decreases with increasing N. In other words, supersymmetry-breaking metastable states become *marginal* (more precisely, marginally unstable) in the thermodynamic limit, having at least one zero mode. This means that even if for any large but finite N there are exponentially numerous stable states, an infinitesimal $O(1/N)$ external field may destabilize some of them, causing the merging of a minimum–saddle pair and making the states disappear. On the other hand, virtual states, i.e. inflection points of the free energy with a very small second derivative, may be stabilized by the field, giving rise to pairs of new states. In other words, the structure of metastable states is fragile and extremely sensitive to external perturbations.

In such a situation we must reconsider the validity of Eq. (7.41). Even if we expect the fluctuation–dissipation relation to always hold inside each given state, the same may not be true when averages over the states are considered, since the number of metastable states can vary dramatically when a small field is applied. If we consider all the TAP states with free energy density f, the average magnetization can be written as

$$\langle m_i \rangle = \frac{1}{\mathcal{N}(f,h)} \sum_\alpha m_i^\alpha, \tag{7.42}$$

where $\mathcal{N}(f,h)$ is the number of states with free energy density f, in the presence of a magnetic field h. Therefore, at the l.h.s. of Eq. (7.41) we differentiate with respect to an external field a sum over all metastable states. The problem is that, due to marginality, some elements in this sum may disappear or appear as the field goes to zero. More precisely, we have,

$$\begin{aligned} \left.\frac{\mathrm{d}\langle m_i\rangle}{\mathrm{d}h_j}\right|_{h=0} &= \lim_{h\to 0}\frac{1}{h_j}\left\{\frac{1}{\mathcal{N}(f,h)}\sum_\alpha m_i^\alpha(h) - \frac{1}{\mathcal{N}(f,0)}\sum_\alpha m_i^\alpha(0)\right\} \\ &\neq \lim_{h\to 0}\frac{1}{\mathcal{N}(f)}\sum_\alpha \frac{1}{h_j}\{m_i^\alpha(h) - m_i^\alpha(0)\} = \langle \partial_i\partial_j F_{\mathrm{TAP}}\{m_i\}^{-1}\rangle. \end{aligned} \tag{7.43}$$

The key point is that the elements in the two summations of (7.43) may be different, because of the action of the field. Therefore, an anomalous contribution arises due to the instability of the whole structure with respect to the field. The fluctuation–dissipation relation (7.41) is thus *violated in supersymmetry-breaking systems.*

7.6 Summary

In this chapter we have introduced a different approach to the statics of the SK model, the TAP approach, which is based on mean field-like equations for the local magnetizations. A crucial quantity within this approach is the 'complexity', which measures the entropy related to the number of metastable states (TAP solutions). Disordered systems like the SK model are characterized by the presence of an exponentially large number of metastable states, which results in a nontrivial spectrum of the complexity as a function of the free energy density of the states. We have looked in some detail at the computation of the complexity and found that:

- The computation of the complexity of the TAP solutions exhibits at a formal level a peculiar form of the BRST supersymmetry.
- The Ward–Takahashi Identities generated by this supersymmetry represent necessary conditions to prove the equivalence between the TAP computations of equilibrium quantities and the standard thermodynamic approach.
- These Ward–Takahashi Identities have a deep physical interpretation corresponding to an average static fluctuation–dissipation relation.

- In the SK model metastable states with high free energy density violate supersymmetry. This is related to the individual nature of these states which at finite N are stable but very close to a nearby saddle, and become marginally unstable in the thermodynamic limit. As a consequence, the overall structure of states is extremely fragile: even small $O(1/N)$ external perturbations may create or destroy many minimum–saddle pairs, changing the number of states dramatically. In this way, averages over TAP states change in a nontrivial way when the external parameters are changed, and the average fluctuation–dissipation theorem does not hold.

References

Annibale A., Cavagna A., Giardina I. and Parisi G. (2003a). *Phys. Rev. E*, **68**, 061103.
Annibale A., Cavagna A., Giardina I., Trevigne E. and Parisi G. (2003b). *J. Phys. A*, **36**, 10937.
Aspelmeier T., Bray A. J. and Moore M. A. (2004). *Phys. Rev. Lett.*, **92**, 087203.
Becchi C., Rouet R. and Stora A. (1975). *Comm. Math. Phys.*, **42**. 127.
Bray A. J. and Moore M. A. (1979). *J. Phys. C*, **12**, L441.
Bray A. J. and Moore M. A. (1980). *J. Phys. C*, **13**, L469.
Cavagna A., Garrahan J. P. and Giardina I. (1998). *J. Phys. A*, **32**, 711.
Cavagna A., Giardina I., Parisi G. and Mézard M. (2003). *J. Phys. A*, **36**, 1175.
Cavagna A., Giardina I. and Parisi G. (2004). *Phys. Rev. Lett.*, **92**, 120603.
Crisanti A., Leuzzi L., Parisi G. and Rizzo T. (2003). *Phys. Rev. B*, **68**, 174401.
Crisanti A., Leuzzi L., Parisi G. and Rizzo T. (2004). *Phys. Rev. Lett.*, **92**, 127203.
De Dominicis C. (1980). *Phys. Rep.*, **607**, 37.
De Dominicis C. and Young A. P. (1983). *J. Phys. A*, **16**, 2063.
Georges A. and Yedidia J. (1991). *J. Phys. A*, **24**, 2173.
Kurchan J. (1991). *J. Phys. A*, **24**, 4969.
Mézard M., Parisi G. and Virasoro M. A. (1986). *Europhys. Lett.*, **1**, 77.
Mézard M., Parisi G. and Virasoro M. A. (1987). *Spin Glass Theory and Beyond*, Singapore, World Scientific, Vol. 9.
Parisi G. (2006). In les Houches Session LX, 2005, *Mathematical Statistical Physics*, ed. A. Bovier, F. Dunlop, A. Van Enter, F. den Hollander and J. Dalibard, Amsterdam, Elsevier Science.
Parisi G. and Rizzo T. (2004). *J. Phys. A*, **37**, 7979.
Plefka T. (1982). *J. Phys. A*, **15**, 1971.
Thouless D. J., Anderson P. W. and Palmer R. (1977). *Phil. Mag.*, **35**, 593.
Tyutin I. V. (1975). Lebedev preprint FIAN 39.
Zinn-Justin J. (1989). *Quantum Field Theory and Critical Phenomena*, Oxford, Clarendon Press.

8

Spin glass above $D = 6$

In the previous chapters we have analyzed in detail the SK model and its solution within the replica method and via mean field-like TAP equations. The physical scenario unveiled by the RSB solution is novel and intriguing, depicting a low temperature phase where ergodicity is broken in a multiplicity of pure states with a nontrivial structure. Yet, so far we have been dealing with a mean field model and one may wonder whether all these surprising results are just an artefact of the long range interaction. This question is indeed very much debated and different points of view exist with conflicting conclusions. In this book hereafter we shall embrace what seems a most natural approach for the finite dimensional model, developing a field theory that has as mean field limit the SK solution described in Chapter 6, and building up a perturbation expansion around it. This is justified if we assume that the physics of the SK model remains qualitatively relevant for a finite dimensional system. Vice versa, we may say that, if we are able to build a well defined field theory and control its perturbation expansion, this is a strong indication of the physical relevance of its content.

Even if our program is conceptually standard routine in field theory, from a practical point of view it is far from simple, given the complex nature of the order parameter. As we shall see, even the analysis of Gaussian fluctuations becomes cumbersome. In this chapter we make the first steps in this program. In Section 8.1 we write the spin glass Lagrangian for a system in finite dimension close to the critical temperature. In Section 8.2 we look at the Gaussian fluctuations and compute the free propagators for a replica symmetric saddle point. In Section 8.3 we consider fluctuations around an RSB saddle point; we also analyze the behaviour of the propagators and the first loop corrections to the equation of state for dimensions $6 < D < 8$. Finally, in the last section we discuss the physical meaning of the propagators.

8.1 The spin glass Lagrangian

Let us consider the Edwards–Anderson model (Edwards and Anderson, 1975).

$$\mathcal{H} = -\sum_{(ij)} J_{ij} S_i S_j - h \sum_i S_i, \tag{8.1}$$

where, now, the sum runs over pairs of neighbours on a D-dimensional lattice. The random couplings are drawn from a Gaussian distribution having zero mean and variance J^2/z, where $z = 2D$ is the number of neighbours. One can then proceed as in Chapter 6 (see the footnotes), to deal with disorder averages and obtain a replicated effective Lagrangian analogous to (6.9). However, so far we have used the same symbol Q^{ab} for the field and its average value (as it turned out, the value of the field at the saddle point). In the following it will be necessary to clearly distinguish between them. So, from now on, we use ϕ_i^{ab} for the field and keep Q_i^{ab} (or Q^{ab}) for its average value. Thus we have

$$\overline{Z^n} = \int \prod_{\substack{(ab)\\ i}} \frac{d\phi_i^{ab}}{\sqrt{2\pi}} \exp\left\{\mathcal{L}\left\{\phi_i^{ab}\right\}\right\}, \tag{8.2}$$

$$\mathcal{L}\left\{\phi_i^{ab}\right\} = -\frac{(\beta J)^2}{2} z \sum_{i,j} \sum_{(ab)} \phi_i^{ab} (K^{-1})_{ij} \phi_j^{ab} + \frac{Nn}{4} (\beta J)^2 + \sum_i \ln \operatorname*{tr}_{\{S_i^a\}} \exp\left\{(\beta J)^2 \sum_{(ab)} \phi_i^{ab} S_i^a S_i^b + \beta h \sum_a S_i^a\right\}. \tag{8.3}$$

Here $K_{ij} = 1$ for nearest neighbour sites and 0 otherwise and, due to the space dependence of the interactions, a distinct field overlap ϕ_i^{ab} must be introduced for any given site i.

In space Fourier Transform $\tilde{K}(k) = \sum_{\mu=1}^{2D} \cos k_\mu a$, where a is the lattice spacing. So that, for small ka, the quadratic term in (8.3) becomes

$$-\frac{1}{2} (\beta J)^2 \sum_{(ab)} \sum_k \phi^{ab}(k) \frac{z}{z - (ka)^2} \phi^{ab}(-k). \tag{8.4}$$

Expanding in $(ka)^2$ one recovers the kinetic energy term in p^2, where $p = ka$ is now the momentum to be integrated up to the ultraviolet (UV) cut off $\Lambda \sim 1/a$. For convenience we may sometimes leave $\sum_p$ to mean integration over the momenta, i.e. $\int^\Lambda d^D p/(2\pi)^D$. Otherwise, close to the critical temperature one can expand in ϕ_i^{ab} as we did in Section 6.3.2 and obtain, after appropriate rescaling, a truncated

Lagrangian analogous to (6.58) (see Bray and Moore 1979):

$$\mathcal{L} = -\frac{1}{2} \sum_p \sum_{(ab)} p^2 (\phi^{ab}(p))^2$$
$$+ \sum_i \left\{ \frac{1}{2} \tau \operatorname{tr} \phi_i^2 + \frac{w}{6} \operatorname{tr} \phi_i^3 + \frac{u}{12} \sum_{ab} \left(\phi_i^{ab}\right)^4 + \frac{1}{2} h^2 \sum_{ab} \phi_i^{ab} \right\}, \quad (8.5)$$

where, we recall, τ is the reduced temperature measured relative to the mean field critical temperature and we have kept only the quartic term which is responsible for RSB at the mean field level. The saddle point (at $p = 0$, with $\phi_i^{ab} = Q^{ab}$), gives back Eq. (6.78):

$$0 = 2\tau Q^{ab} + w \sum_c Q^{ac} Q^{cb} + \frac{2u}{3} (Q^{ab})^3 + h^2. \quad (8.6)$$

We have already studied the solutions of (8.6) in Chapter 6:

(i) above the AT line $Q^{ab} = Q$ (RS solution), with $Q = 0$ in zero field;
(ii) below the AT line $Q^{ab} = Q(x)$ (RSB solution), where x is the replica overlap $a \cap b$.

8.2 RS propagators

In Chapter 6 we have also studied the fluctuation Hessian *around the RS solution* Q. In fact we have (quasi) diagonalized the quadratic form for the fluctuation field via Replica Fourier Transforms (RFT), see Eq. (6.49). In the present case, the analysis of fluctuations proceeds exactly in the same way, the only difference being that now the fluctuation field acquires a space dependence:

$$\phi_i^{ab} = Q + \varphi_i^{ab}. \quad (8.7)$$

Thus, the contribution of Gaussian fluctuations to the spin glass Lagrangian reads

$$\mathcal{L}^{(2)} = -\frac{1}{2} \sum_p \left\{ \frac{1}{2} M_{\mathrm{R}} \sum_{\hat{a}'=1}^{n-1} (\Phi_{\hat{a}';-\hat{a}'}(p))^2 + M_{\mathrm{L}} (\varphi_{\hat{0};\hat{0}}(p))^2 \frac{n}{2(n-1)} \right.$$
$$\left. + \sum_{\hat{t}'=1}^{n-1} \left[\frac{1}{2} M_{\mathrm{R}} \sum_{\hat{a}''=1}^{n-2} |\Phi_{\hat{a}'';\hat{t}'-\hat{a}''}(p)|^2 + M_{\mathrm{A}} |\varphi_{\hat{0};\hat{t}'}(p)|^2 \frac{n}{n-2} \right] \right\}; \quad (8.8)$$

here $\hat{a}', \hat{t}' \neq 0$, $\hat{a}'' \neq 0, \hat{t}'$.

As in (6.49), we have

$$\begin{cases} M_{\mathrm{R}} = M_1 - 2M_2 + M_3, \\ M_{\mathrm{A}} = M_1 - 4M_2 + 3M_3 + n(M_2 - M_3), \\ M_{\mathrm{L}} = M_{\mathrm{A}} + n(M_2 - M_3) + \dfrac{n(n-1)}{2} M_3. \end{cases} \tag{8.9}$$

But now $M_1 = p^2 - 2(\tau + uQ^2)$ includes also the kinetic term in p^2. Hence, the replicon eigenvalue acquires the form

$$M_{\mathrm{R}} = p^2 - \frac{4u}{3} Q^2 + \frac{h^2}{Q}. \tag{8.10}$$

If we recall the scaling with the reduced temperature (see Section 6.3.2), the replicon mode has then a *small (negative) mass* which scales as $\sim \tau^2$ (for $h = 0$). On the other hand, the longitudinal and anomalous modes behave in the $n \to 0$ limit as

$$M_{\mathrm{L}} = M_{\mathrm{A}} = M_{\mathrm{R}} + 2wQ \tag{8.11}$$

with a *large (positive) mass* $\sim \tau$.

Up to now, we have been able to compute the eigenvalues of the Hessian matrix. Propagators, i.e. fluctuation correlations, are then easily obtained by inversion of the (quasi) diagonalized form of the Hessian, Eq. (8.8). They can be directly read off the quadratic form. We have, for the longitudinal and anomalous propagators in the RFT space,

$$\begin{aligned} G^{\mathrm{L}}_{\hat{0};\hat{0}} &\equiv \langle \varphi_{\hat{0};\hat{0}}\, \varphi_{\hat{0};\hat{0}} \rangle \frac{n}{2(n-1)} = \frac{1}{M_{\mathrm{L}}}, \\ G^{\mathrm{A}}_{\hat{r}';\hat{s}'} &\equiv \langle \varphi_{\hat{0};\hat{r}'}\, \varphi^*_{\hat{0};\hat{s}'} \rangle \frac{n}{n-2} = \delta_{\hat{r}';\hat{s}'} \frac{1}{M_{\mathrm{A}}}. \end{aligned} \tag{8.12}$$

Note that the total momentum (in RFT space) is *null*. For the RFT replicon propagators we have

$$G^{\mathrm{R}}_{\hat{a}';\hat{b}'} \equiv \langle \Phi_{\hat{a}';-\hat{a}'}\, \Phi_{\hat{b}';-\hat{b}'} \rangle = \left[\delta_{\hat{a}';\hat{b}'} + \delta_{\hat{a}';-\hat{b}'} - \frac{2}{n-1} \right] \frac{1}{M_{\mathrm{R}}}. \tag{8.13}$$

In this case the propagator is not exactly diagonalized because we have the constraint on Φ, viz. $\sum_{\hat{a}} \Phi_{\hat{a}';-\hat{a}'} = 0$. We could have made one more RFT, now mod $n-1$, and formally diagonalized it. It turns out that it is just as simple to leave it as it is. Likewise,

$$G^{\mathrm{R}}_{\hat{a}'';\hat{r}'-\hat{b}''} = \langle \Phi_{\hat{a}'';\hat{r}'-\hat{a}''}\, \Phi^*_{\hat{b}'';\hat{r}'-\hat{b}''} \rangle = \left[\delta_{\hat{a}'';\hat{r}'} + \delta_{\hat{a}'';\hat{r}'-\hat{b}''} - \frac{2}{n-2} \right] \frac{1}{M_{\mathrm{R}}}. \tag{8.14}$$

From them we can easily obtain propagators in the direct replica space, just by exploiting their RFT expression, e.g.

$$G_1 \equiv \langle \varphi_{ab}\varphi_{ab}\rangle = \frac{1}{n(n-1)}\sum_{ab}\langle \varphi_{ab}\varphi_{ab}\rangle$$
$$= \frac{1}{n\,(n-1)}\sum_{\hat{a}\hat{b}}\langle \varphi_{\hat{a}\hat{b}}\varphi_{-\hat{a}-\hat{b}}\rangle, \tag{8.15}$$

$$G_2 \equiv \langle \varphi_{ab}\varphi_{ac}\rangle = \frac{1}{n(n-1)(n-2)}\left\{\sum_{abc}\left(1-\delta_{b;c}\right)\langle \varphi_{ab}\varphi_{ac}\rangle\right\}$$
$$= \cdots, \tag{8.16}$$

$$G_3 \equiv \langle \varphi_{ab}\varphi_{cd}\rangle = \frac{1}{n(n-1)(n-2)(n-3)}$$
$$\times\left\{\sum_{abcd}(1-\delta_{a;c}-\delta_{a;d}-\delta_{b;c}-\delta_{b;d}+\delta_{a;c}\delta_{b;d}+\delta_{a;d}\delta_{b;c})\langle \varphi_{ab}\varphi_{cd}\rangle\right\}$$
$$= \cdots \tag{8.17}$$

In particular, the combinations of propagators corresponding to the replicon and longitudinal–anomalous sub-spaces read respectively

$$G^{\mathrm{R}} \equiv G_1 - 2G_2 + G_3 = \frac{1}{M_{\mathrm{R}}}, \tag{8.18}$$

$$G^{\mathrm{A}} = G^{\mathrm{L}} \equiv G_1 - 4G_2 + 3G_3 = \frac{1}{M_{\mathrm{L}}}. \tag{8.19}$$

The fact that G^{R} gets a *pole on the real positive axis*, below the AT line, forces the search for an alternative saddle point to the RS one (with RSB).

For completeness we quote the result for the Gs (first written by Pytte and Rudnick, 1979):

$$G^{ab;ab} \equiv G_1 = \frac{1}{M_{\mathrm{R}}}(1+X+X^2), \tag{8.20}$$

$$G^{ab;ac} \equiv G_2 = \frac{1}{M_{\mathrm{R}}}\left(\frac{X}{2}+X^2\right), \tag{8.21}$$

$$G^{ab;cd} \equiv G_3 = \frac{1}{M_{\mathrm{R}}}(X^2), \tag{8.22}$$

with

$$X \equiv -\frac{2\,(M_2-M_3)}{M_{\mathrm{R}}-2(M_2-M_3)}. \tag{8.23}$$

8.3 RSB propagators

Below the AT line, the RS saddle point is unstable with respect to Gaussian fluctuations and, at the mean field level, one must consider an RSB saddle point, as discussed in detail in Chapter 6. Let us then consider again the Lagrangian (8.5) and expand it around a generic (not necessarily RS) mean field solution Q^{ab}, satisfying the stationarity equation (8.6). From (8.5), under replacement of the field ϕ_i^{ab} by

$$\phi_i^{ab} = Q^{ab} + \varphi_i^{ab}, \tag{8.24}$$

we get

$$\begin{aligned}\mathcal{L}\{\phi_i^{ab}\} = \mathcal{L}^{(0)}\{Q^{ab}\} - \frac{1}{2}\sum_{(ab)}\sum_{(cd)}\sum_{p}\varphi^{ab}(p)\,M^{ab;cd}(p)\varphi^{cd}(-p) \\ + \frac{w}{6}\sum_i \operatorname{tr}\varphi_i^3 + \frac{u}{3}\sum_{a,b} Q^{ab}\sum_i \left(\varphi_i^{ab}\right)^3 + \frac{u}{12}\sum_{a,b}\sum_i \left(\varphi_i^{ab}\right)^4,\end{aligned} \tag{8.25}$$

where, now

$$\begin{cases} M_1 = M^{ab;ab}(p) = p^2 - 2(\tau + u(Q^{ab})^2), & \\ M_2 = M^{ab;ac} \quad = -wQ^{bc}, & b \neq c, \\ M_3 = M^{ab;cd} \quad = 0, & (ab) \neq (cd). \end{cases} \tag{8.26}$$

The precise form of the Hessian depends on the structure of the overlap matrix Q^{ab}. In the RS case discussed in the previous section, the diagonalization of $M^{ab;cd}$ led to a large longitudinal–anomalous eigenvalue (the large mass $w\,Q \sim \tau$) and a small replicon eigenvalue (the small mass $-(4/3)u\,Q^2 \sim -\tau^2$ in zero field, negative under the AT line). We have seen in Chapter 6 the improvement achieved under the one-step RSB, that reduces the negative value of the entropy and the negativeness of the replicon small mass. What we would like to do now is to diagonalize the Hessian for the Parisi solution of Eqs. (6.83), (6.85) and (6.89) to see whether and how the instability is cured by the full RSB solution. Then we can compute the free propagators by inversion and set up a perturbative expansion around the mean field solution.

The Hessian diagonalization can be done, as for the RS solution, by resorting to the RFT technique. The final result is that under the full RSB Parisi scheme the two eigenvalues found for the RS solution, the replicon one and the longitudinal–anomalous one, become two continuous bands, as shown in Fig. 8.1. In particular, the replicon mass becomes a band bounded below by zero (De Dominicis and Kondor, 1983). Under an increasing magnetic field, each band shrinks to a point as one crosses the AT line.

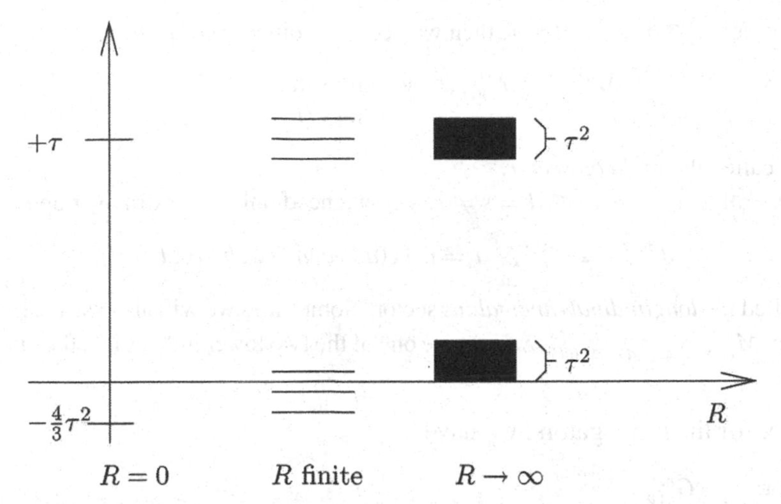

Figure 8.1 Structure of the replicon eigenvalues with progressive steps R of Replica Symmetry Breaking

Once the eigenvalues are obtained, one wants to invert the matrix M and get the free propagators $G^{ab;cd}$. They are given by the 'unitarity' equation

$$\sum_{(cd)} M^{ab;cd}\, G^{cd;ef} = \delta_{(ab);(ef)} = \frac{1}{2}(\delta_{a;e}\,\delta_{b;f} + \delta_{a;f}\,\delta_{b;e}) \tag{8.27}$$

Having solved for G, one can then expand (8.25) in w, u and get the loop corrections using a Wick theorem, as in standard field theory. This is easy to say, but rather difficult to carry out. In this chapter we take the first steps in this direction. First we describe how one can conveniently parametrize the Hessian matrix and the propagators, then derive the equations satisfied by the Gs, and hint at solutions in some appropriate range. The full solution will be obtained in the next chapters.

8.3.1 Parametrization

Let us first remain within a discrete RSB ansatz with R steps. Then

$$Q^{ab} = Q_r, \tag{8.28}$$

r being the replica overlap $a \cap b$ (i.e. the co-distance). For the propagators (or their inverse, the M matrix) we have four replicas (i.e. generically three overlaps). Recalling the ultrametric structure of the RSB solution, we thus have two possibilities for the matrix $M^{ab;cd}$:

- If generically $a \cap b \equiv c \cap d \equiv r$, then we need two other overlaps u, v:

$$M^{ab;cd} \equiv M^{r;r}_{u;v}, \qquad \begin{aligned} u &= \max(a \cap c,\ a \cap d), \\ v &= \max(b \cap c,\ b \cap d). \end{aligned} \tag{8.29}$$

This is called the *replicon sector*.
- If generically $a \cap b = r$, $c \cap d = s \neq r$, then we need only *one* extra overlap t:

$$M^{ab;cd} \equiv M^{r;s}_{t}, \quad t = \max(a \cap c, a \cap d, b \cap c, b \cap d), \tag{8.30}$$

also called the *longitudinal–anomalous* sector. Sometimes we will also use the redundant notation $M^{r;s}_{\max(a\cap c, a\cap d);\max(b\cap c, b\cap d)}$, where one of the two lower indices is t, the other being r or s.

Likewise, for the propagators, we have

- R-sector: $G^{r;r}_{u;v}$,
- L-A-sector: $G^{r;s}_{t}$.

In the limit $R \to \infty$ of continuous RSB the parametrization indices assume values in the interval [0, 1].

8.4 The unitarity equation in the near infrared

Knowing now the Hessian matrix $M^{ab;cd}$ as of (8.26), we want to use the unitarity equation (8.27) to obtain the set of (bare) propagators. As anticipated, this is far from simple, contrary to standard field theory where this inversion is usually trivial. For example, let us write the first of these equations:

$$\sum_{(cd)} M^{ab;cd}\, G^{cd;ab} = 1, \tag{8.31}$$

that is

$$M^{ab;ab}\, G^{ab;ab} + 2 \sum_{c \neq a,b} M^{ab;ac}\, G^{ac;ab} = 1, \tag{8.32}$$

since $M^{ab;cd} = 0$ when $c, d \neq a, b$.

Using Eq. (8.26) and the parametrization for the propagator described in the previous section, we can compute the sum $\sum_c$ in a way analogous to the calculation of $\sum_c Q^{ac} Q^{cb}$ in Section 6.4.2. In this case there are four distinct contributions corresponding to the four tree structures illustrated in Fig. 8.2. Setting $a \cap b = x$ they read, respectively:

- case (A) $b \cap c = t$; $t < x$,

$$-w \sum_{t<x} (p_t - p_{t+1}) Q_t G^{t;x}_{1;t} \longrightarrow w \int_0^x \mathrm{d}t\ Q(t) G^{t;x}_{1;t}; \tag{8.33}$$

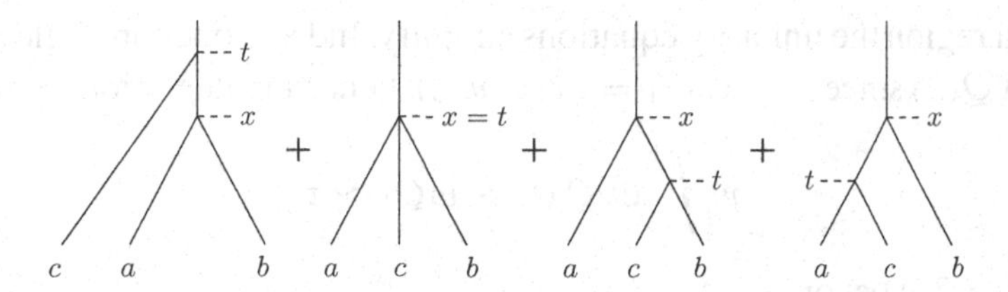

Figure 8.2 Tree structures for the cases (A), (B), (C), (D) discussed in the text

- case (B) $b \cap c = t; t = x$,

$$-w(p_x - 2p_{x+1})Q_x G_{1;x}^{x;x} \longrightarrow wxQ(x)G_{1;x}^{x;x}; \tag{8.34}$$

- case (C) $b \cap c = t; t > x$,

$$-w\sum_{t>x}(p_t - p_{t+1})Q_t G_{1;t}^{x;x} \longrightarrow w\int_x^1 \mathrm{d}t\ Q(t)G_{1;t}^{x;x}; \tag{8.35}$$

- case (D) $a \cap c = t; t > x$,

$$-w\sum_{t>x}(p_t - p_{t+1})Q_t G_{1;t}^{t;x} \longrightarrow w\int_x^1 \mathrm{d}t\ Q(t)G_{1;t}^{t;x}. \tag{8.36}$$

Where we have kept the same notation (t, x) for the discrete (l.h.s) and continuous (r.h.s) version of (8.33)–(8.36). Altogether, directly in the $R \to \infty$ limit, we get

$$[p^2 - 2(\tau + uQ^2(x))]G_{1;1}^{x;x} + 2w\left[\int_0^x \mathrm{d}t\ Q\,(t)\,G_{1;x}^{t;x} + x\ Q\,(x)\,G_{1;x}^{x;x}\right.$$
$$\left. + \int_x^1 \mathrm{d}t\ Q\,(t)\,G_{1;t}^{x;x} + Q\,(x)\int_x^1 \mathrm{d}t\ G_{1;t}^{t;x}\right] = 1. \tag{8.37}$$

There are nine such coupled equations determining the bare propagators G. It is rather surprising that these equations can be solved exactly. In this chapter we indeed get an exact solution for the most dangerous sector (the replicon sector). For the time being, let us focus on the so-called *near infrared* regime, that is for momenta that may be comparable to the large mass, but are much larger than the small one:

$$p^2 \gg \tau^2. \tag{8.38}$$

As we shall see, this regime gives an accurate description of the propagators if we consider dimensions above the upper critical dimension (for a cubic coupling) $D_c = 6$. In this case we are not in a region dominated by infrared singularities (i.e. $p^2 \ll \tau^2$) but dominant contributions come from $p^2 \gg \tau^2$, including $p^2 \sim \tau$ (the large mass).

In such a region the unitarity equations simplify. Indeed, one can neglect terms in $Q^2(x)$, or $x\,Q(x)$ since $x < x_1$ ($x_1 = 2\tau(u/w^2)$), hence in integrations we have, e.g.

$$w \int_x^1 \mathrm{d}t\, Q(t) \sim wQ_1 \sim \tau. \tag{8.39}$$

As a result (8.37) becomes

$$(p^2 - 2wQ_1)G^{x;x}_{1;1} + 2wQ(x)\,G^{x_1;x}_{1;x_1} + 2wQ_1 G^{x;x}_{1;x_1} = 1, \tag{8.40}$$

now a linear *algebraic* equation. There are nine such equations that are now easily solved for the Gs, describing the *near infrared* sector ($p^2 \gg \tau^2$, i.e. $p^2 \gtrsim \tau$). We will not display here the complete list of the near infrared equations and the corresponding solution for the propagators (see De Dominicis *et al.*, 1998), but just add a few comments. First, one finds that in the near infrared the propagators describing correlations within a single ergodic phase, which are obtained by setting $x = y = x_1$ in $G^{x,y}_{z_1,z_2}$, have the same form as the propagators around an RS mean field saddle point. Second, all the propagator components have two poles, one at $p = 0$ and one at $p^2 = -2wq_1$ (we recall that we are in the $p^2 \gg \tau^2$ limit). These singularities become two continuous bands in the exact result for the bare propagators.

8.5 One-loop correction to the equation of state

In the near infrared regime one can easily write the first one-loop correction to Eq. (8.6) due to the w and u coupling appearing in the Lagrangian. The equation of state can be viewed as requiring local equilibrium for the homogeneous solution Q^{ab}, or in other terms, requiring that the average value of the fluctuation $\langle \varphi^{ab} \rangle$ be zero. To zero order this is equivalent to Eq. (8.6). Higher orders are obtained using the complete measure (8.25) to compute the average and expanding with respect to the interaction terms in u and w. By means of the Wick theorem, to one loop we get

$$\begin{aligned} &2\tau\, Q^{ab} + w \sum_c Q^{ac}\, Q^{cb} + \frac{2u}{3}(Q^{ab})^3 \\ &\quad + \frac{1}{z} \sum_p \left\{ w \sum_c G^{ac;cb}(p) + 2u\, Q^{ab}\; G^{ab;ab}(p) \right\} + O(1/z^2) = 0. \end{aligned} \tag{8.41}$$

That is, in the continuum limit,

$$\begin{aligned} &2\left[\tau Q(x) + \frac{u}{3} Q^3(x)\right] - w\left[\int_0^x \mathrm{d}t\, Q^2(t) + x\,Q^2(x) + 2Q(x)\int_x^1 \mathrm{d}t\; Q(t)\right] \\ &\quad + \frac{1}{z}\sum_p 2u\,Q(x) G^{x;x}_{1;1}(p) - \frac{w}{z}\sum_p \left\{ \int_0^x \mathrm{d}t\; G^{t;t}_{1;1} + x\, G^{x;x}_{1;1} + 2\int_x^1 \mathrm{d}t\; G^{x;t}_{1;x} \right\} \\ &\quad + \cdots = 0. \end{aligned} \tag{8.42}$$

For the region $p^2 \gg \tau^2$, dropping terms in τ^2, one can reduce the term in w/z to

$$-2\frac{w}{z}\sum_p G_{1;x}^{x;x_1}. \tag{8.43}$$

Using the equations briefly described above for the near infrared sector, we get an explicit contribution from (8.42) in $Q(x)$ and in $Q^3(x)$. This last contribution thus renormalizes (to one loop) the u coupling into

$$\tilde{u} = u + 12\frac{w^4}{z}\int \frac{\mathrm{d}^D p}{(2\pi)^D}\frac{1}{p^4(p^2+2wQ_1)^2}. \tag{8.44}$$

For $D > 8$, $\tilde{u}$ is only numerically different from u. A similar argument also holds for the other coupling constants, namely w and the reduced temperature τ. As a consequence, for $D > 8$ the equation of state is only slightly corrected. In particular, it is found that the correction goes in the direction of enhancing RSB effects (for example, the slope of $Q(x)$ decreases, while the breakpoint x_1 increases with decreasing D). The same result is found for the renormalized masses, indicating that mean field theory as described so far is perturbatively stable above $D = 8$.

The situation is different for $6 < D < 8$. In this case $\tilde{u}$ is qualitatively changed (recall that $2wQ_1 \sim \tau$), viz.

$$\tilde{u} = u + \frac{cw^4}{z\tau^{(8-D)/2}}, \tag{8.45}$$

and for temperatures close to T_c, one can neglect u, and write

$$\tilde{u} \sim 1/(\tau)^{(8-D)/2}. \tag{8.46}$$

The ordinary loop expansion then breaks down and something has to be done to cure this situation. Note that since the bare u coupling becomes negligible below $D = 8$ we could have used as a starting point the effective Lagrangian for the fluctuation field φ^{ab} without the quartic interaction term. This is, however, not possible for the equation of state itself since the quartic term is crucial to determine the RSB structure of the mean field order parameter Q^{ab}.

The presence of such a singularity does not mean that $D = 8$ should be considered as the upper critical dimension. The logarithms appearing in $D = 8$ are not going to proliferate, which can be confirmed by power counting. The singularity in $(\tau)^{(D-8)/2}$ comes from the one-loop correction to the coupling u, i.e. to the four-point function (the fourth derivative with respect to $Q(x)$ of the free energy) at zero external momenta – see Fig 8.3 (right). This is the analogue of a singularity $(\tau)^{(D-6)/2}$ that occurs in the pure ϕ^4 theory for the six-point function at zero external momenta (Fig. 8.3 (left)). It is an isolated singularity which does not affect other multi-point functions. Indeed, the corresponding graph, as an insertion, always appears at nonzero momenta and its singularity is smoothed out by the loop

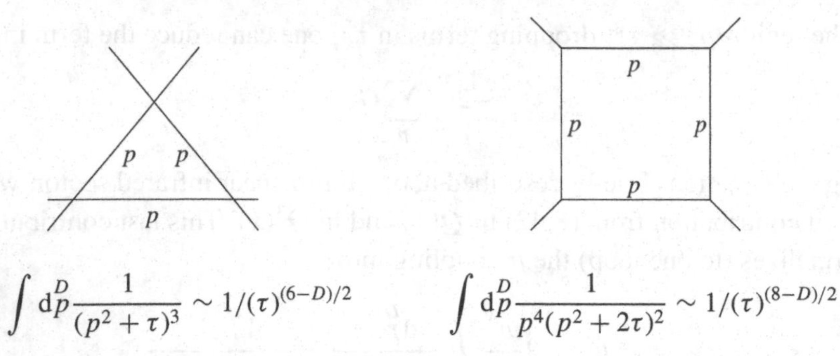

p p
p
p
p p
p
$\int d^D p \frac{1}{(p^2+\tau)^3} \sim 1/(\tau)^{(6-D)/2}$
$\int d^D p \frac{1}{p^4(p^2+2\tau)^2} \sim 1/(\tau)^{(8-D)/2}$

Figure 8.3 Graphs and corresponding singularities for a ϕ^4 (left) and ϕ^3 (right) field theory

integrations. In the standard ϕ^4 theory the six-point function does not enter physically relevant quantities and is therefore of small interest *per se*. The same is not true in the spin glass case where, as we have seen, the four-point function at zero external momenta directly appears in the equation of state and determines the shape of the order parameter $Q(x)$. So what is to be done is to take into account the singular behaviour of $\tilde{u}$ and modify accordingly mean field theory down to $D = D_c = 6$. Taking (8.45) and (8.46) into account, one obtains now

$$\begin{cases} wQ(x) = \dfrac{w^2}{2\tilde{u}}x & = \dfrac{(\tau)^{(8-D)/2}}{2(w^2/z)}x, \\ \quad x_1 = \dfrac{w^2}{z}\tau^{1-(8-D)/2} & = \dfrac{w^2}{z}\tau^{(D-6)/2}. \end{cases} \tag{8.47}$$

As a consequence the large mass remains unchanged, while the small mass acquires a nontrivial scaling:

$$M = wQ_1 \sim \tau,$$
$$m = \tilde{u}Q_1^2 \sim \tau w^2 \left(\frac{\tau^{(D-6)/2}}{z}\right). \tag{8.48}$$

The two masses' functional dependence in τ becomes identical as $D \to D_c = 6$ (hence standard scaling with a single mass is recovered, as was first noticed by Fisher and Sompolinsky, 1985). The breakpoint loses any τ dependence in the same limit. What this all amounts to is that, upon approaching $D = 6$ from above, $Q(x)$ becomes of the form

$$Q(x) = \tau^\beta f(x), \tag{8.49}$$

where $\beta = 1$ and $f(x)$ is now temperature independent. Likewise, it is easy to show that there is a common mass scale recovered as $D \to 6+$ which is characterized by a single exponent $\nu = 1/2$; still there remains a w^2/z factor to mark the separation between the two scales. Anticipating what is happening below $D = 6$, one has β

acquiring loop corrections $\beta = 1 + \varepsilon/2 + O(\varepsilon^2)$ (with $\varepsilon = 6 - D$) and the mass ratio w^2/z becoming of order $\varepsilon + O(\varepsilon^2)$ (Harris, Lubensky and Chen, 1976; see Appendix B), with no guarantee that the two bands won't mix as ε is increasing.

To conclude this section we note that in standard systems the upper critical dimension D_c ($D_c = 4$ for ϕ^4 systems) is at the same time the dimension above which mean field ($z \to \infty$) is valid, and the dimension below which fluctuations dominate. Here there are two dimensions for those two properties. Above $D = 8$ mean field is valid, below $D = 6$ fluctuations take over and infrared divergences arise. In between we have a mean field corrected to take account of effects of the irrelevant dangerous couplings. In this sense, one has to consider $D = 6$ as the upper critical dimension for the spin glass field theory.

To go below $D = 6$, one has now to face the tough problem of solving for the (bare) propagators not only in the near infrared sector ($p^2 \sim M$) but everywhere, including the deep infrared sector ($p^2 \ll m$).

8.6 Physical meaning of the RSB propagators

In the previous sections we have introduced and discussed the propagators $G^{ab;cd}$ and their parametrization. Clearly, these propagators have a direct interpretation within the effective replicated field theory, corresponding to well defined correlation functions between the fluctuation fields φ_i^{ab}. In this section we wish to discuss more extensively the meaning of the propagators and relate them to physical observables such as standard correlation functions and susceptibilities.

8.6.1 The correlation overlaps

We have already discussed elsewhere (see, e.g., Sections 5.2.4 and 6.4.4) the physical interpretation of the spin overlap. If we consider two identical copies a and b of the system, their spin overlap

$$Q^{ab} = \frac{1}{N} \sum_i \overline{\langle S_i \rangle_a \langle S_i \rangle_b} \tag{8.50}$$

can be computed with the replica method and is given by the saddle point value of the corresponding element of the overlap matrix, as it appears in the replicated effective Lagrangian. For example, for the SK model we found (see Eq. (6.11)) $Q^{ab} = \langle S^a S^b \rangle$, the average being computed with the weight (6.12). The overlap as defined in (8.50) can be interpreted as follows. If at low temperature there are many pure states, each one of the two systems a and b will independently sample them. Q^{ab} thus represents the overlap between a pair of such states, the one sampled by system a and the one sampled by system b. Indeed, as we have discussed in

Section 5.2.4, the overlap distribution $P(Q)$ can be identified with the distribution of the overlap between pure states (Parisi, 1983). In this sense, the indices a and b can also be thought of as referring to pure states. For the SK model we found that $Q^{ab} = Q(x)$ is distributed according to the probability $P(Q) = \mathrm{d}x/\mathrm{d}Q$, which relates the spin overlap Q to the replica overlap x.

In the same spirit we may now consider (Temesvari *et al.*, 1988) the *correlation overlap*

$$C^{ab}(r) = \frac{1}{N}\sum_i \overline{\langle S_i S_j\rangle_a \langle S_i S_j\rangle_b}, \tag{8.51}$$

with $r \equiv r_i - r_j$. In terms of the replicated effective Lagrangian (8.3) (or, equivalently, Lagrangian (8.5)) this can be expressed as

$$\begin{aligned} C^{ab}(r) &= \frac{1}{N}\sum_i \langle \phi_i^{ab}\phi_j^{ab}\rangle \\ &= (Q^{ab})^2 + \frac{1}{N}\sum_i \langle \varphi_i^{ab}\varphi_j^{ab}\rangle \\ &= (Q^{ab})^2 + G^{ab;ab}(r), \end{aligned} \tag{8.52}$$

which in Fourier Transform reads

$$C^{ab}(p) = N\delta^{Kr}_{p;0}(Q^{ab})^2 + G^{ab;ab}(p). \tag{8.53}$$

We then find a relationship between the connected part of the correlation overlap, $C^{ab}_{\mathrm{c}}(p) = C^{ab}(p) - N\delta^{Kr}_{p;0}(Q^{ab})^2$, and the propagator:

$$C^{ab}_{\mathrm{c}}(p) = G^{ab;ab}(p) = G^{x;x}_{1;1}(p), \tag{8.54}$$

where $x = x(Q^{ab})$, i.e. $x = a \cap b$. Equation (8.54) gives direct physical meaning to our propagators, showing what the relation is both in replica space and in configuration space of the pure states between which the correlation is calculated.

Likewise one can consider three identical copies a, b and c of the system (or three pure states) and define

$$C^{abc}(r) \equiv \frac{1}{N}\sum_i \overline{\langle S_i S_j\rangle_a \langle S_i\rangle_b \langle S_j\rangle_c}. \tag{8.55}$$

For the Fourier Transform of the connected part we easily find

$$C^{abc}_{\mathrm{c}}(p) = G^{x;y}_{1;z}, \tag{8.56}$$

with

$$\begin{aligned} x &= x(Q^{ab}), \\ y &= y(Q^{ac}), \\ z &= \max[x(Q^{ab}), x(Q^{bc})]. \end{aligned} \tag{8.57}$$

Finally, for the connected part of

$$C^{abcd}(r) = \frac{1}{N}\sum_i \overline{\langle S_i\rangle_a \langle S_i\rangle_b \langle S_j\rangle_c \langle S_j\rangle_d}, \tag{8.58}$$

we get

$$C_{\rm c}^{abcd}(p) = G^{x;y}_{z_1;z_2}(p), \tag{8.59}$$

with

$$\begin{aligned} x &= x(Q^{ab}),\\ y &= y(Q^{cd}),\\ z_1 &= \max[x(Q^{ac}), x(Q^{ad})],\\ z_2 &= \max[x(Q^{bc}), x(Q^{bd})], \end{aligned} \tag{8.60}$$

where only three of the four indices x, y, z_1, z_2 are distinct. Note that a propagator whose overlaps are $\geq x_1$ is the spin correlation overlap *inside a single state.*

In all the above $(1/N\sum_i)$ and the overbar become redundant for $N \to \infty$

8.6.2 Correlation functions

Let us now consider the three correlation functions that are usually defined for a spin glass system (the redundant $(1/N)\sum_i$) is now removed):

$$\begin{aligned} C_1(r) &= \overline{\langle S_i S_j\rangle^2} - \overline{\langle S_i\rangle^2}\,\overline{\langle S_j\rangle^2},\\ C_2(r) &= \overline{\langle S_i S_j\rangle\langle S_i\rangle\langle S_j\rangle} - \overline{\langle S_i\rangle^2}\,\overline{\langle S_j\rangle^2},\\ C_3(r) &= \overline{\langle S_i\rangle^2\langle S_j\rangle^2} - \overline{\langle S_i\rangle^2}\,\overline{\langle S_j\rangle^2}, \end{aligned} \tag{8.61}$$

with, as usual, $r = r_i - r_j$. In particular, two combinations of these correlations turn out to have a deep physical relevance, being related to the spin glass and the nonlinear susceptibilities. These are

$$\begin{aligned} \chi^{\rm SG}(r) &= C_1(r) - 2C_2(r) + C_3(r)\\ &- \overline{[\langle S_i S_j\rangle \quad \langle S_i\rangle\langle S_j\rangle]^2}, \end{aligned} \tag{8.62}$$

$$\begin{aligned} \chi^{\rm NL}(r) &= C_1(r) - 4C_2(r) + 3C_3(r)\\ &= \overline{[\langle S_i S_j\rangle - \langle S_i\rangle\langle S_j\rangle][\langle S_i S_j\rangle - 3\langle S_i\rangle\langle S_j\rangle]}. \end{aligned} \tag{8.63}$$

In terms of the standard susceptibility $\chi_{ij} = \langle S_i S_j \rangle - \langle S_i \rangle \langle S_j \rangle$, Eqs. (8.62) and (8.63) then correspond to

$$\chi^{\mathrm{SG}} = \overline{\chi_{ij}^2}, \tag{8.64}$$

$$\chi^{\mathrm{NL}} = -\frac{1}{2} \overline{\frac{\partial^2}{\partial h_i \partial h_j} \chi_{ij}|_{h=0}}. \tag{8.65}$$

Using the replicated effective field theory, these correlation functions can be expressed in terms of the overlap correlations introduced before, and therefore in terms of the propagators. Consider, for example, the term $\overline{\langle S_i S_j \rangle^2}$. Following the same lines as in Section 6.4.5 we get

$$\begin{aligned}
\overline{\langle S_i S_j \rangle^2} &= \frac{1}{Z^2} \operatorname*{tr}_{\{S_i\}} \operatorname*{tr}_{\{S_i'\}} \overline{\exp\{-\beta \mathcal{H}(S_i) - \beta \mathcal{H}(S_i')\}(S_i S_j)(S_i' S_j')} \\
&= \lim_{n \to 0} \operatorname*{tr}_{\{S_i^a\}} S_i^a S_i^b S_j^a S_j^b \overline{\exp\left\{-\beta \sum_a \mathcal{H}(S_i^a)\right\}} \\
&= \lim_{n \to 0} \left\langle \phi_i^{ab} \phi_j^{ab} \right\rangle,
\end{aligned} \tag{8.66}$$

where the last average is over $e^{\mathcal{L}}$ with $\mathcal{L}$ as in (8.5). Substituting $\phi_i^{ab} = Q^{ab} + \varphi_i^{ab}$ we get

$$\begin{aligned}
\overline{\langle S_i S_j \rangle^2} &= \lim_{n \to 0} \frac{1}{n(n-1)} \sum_{ab} C^{ab}(r) \\
&= \lim_{n \to 0} \frac{1}{n(n-1)} \left[\sum_{ab} (Q^{ab})^2 + \sum_{ab} G^{ab;ab}(r) \right].
\end{aligned} \tag{8.67}$$

Following the same procedure one can compute all the components of Eqs. (8.61) and find explicit expressions for the spin glass and nonlinear correlations (and related susceptibilities). For example, for the spin glass correlation one gets (directly in momentum space)

$$\begin{aligned}
\chi^{\mathrm{SG}}(p) &= \delta_{p;0}^{Kr} \lim_{n \to 0} \frac{1}{3n(n-1)} \left[\sum_{ab} (Q^{ab})^2 + \sum_{abc} Q^{ab} Q^{ac} \right] \\
&+ \lim_{n \to 0} \frac{1}{n(n-1)} \sum_{ab} \left\{ G^{ab;ab}(p) - \frac{2}{(n-2)} \sum_{c \neq b} G^{ab;ac}(p) \right. \\
&\left. + \frac{1}{(n-2)(n-3)} \sum_{c,d \neq a,b} G^{ab;cd}(p) \right\}.
\end{aligned} \tag{8.68}$$

The detailed expression of the correlation functions obviously depends on the replica structure of the mean field solution Q^{ab}. For an RS saddle point we easily find

$$\chi^{\mathrm{SG}}(p) = G^{\mathrm{R}}(p) = \frac{1}{p^2 - \frac{4u}{3}Q^2 + \frac{h^2}{Q}},$$

$$\chi^{\mathrm{NL}}(p) = G^{\mathrm{L}}(p) = \frac{1}{p^2 + 2wQ}, \tag{8.69}$$

that is, fluctuations contributing to the spin glass and nonlinear correlation functions are given, respectively, by the replicon and longitudinal–anomalous propagators. This is the reason why in the literature the correlations (8.62) and (8.63) are often referred to as the replicon and longitudinal correlations. Here we prefer a different notation since this correspondence only holds for an RS ansatz (see Section 9.5). Note that the spin glass susceptibility, given by the $p \to 0$ limit of $\chi^{\mathrm{SG}}(p)$, diverges at the Almeida–Thouless line below which it becomes unstable.

In general, to obtain the correlations within the continuous full RSB ansatz one needs to appropriately parametrize and compute the propagators. In the next chapter we show how this problem can be handled. For the time being let us start to look at the behaviour *inside a given state*. This means that, for example, in (8.68) we do not perform the summations, but fix all overlaps $a \cap b$, $a \cap c$, $c \cap d$ to values larger than x_1. In the near infrared approximation we get

$$\chi^{\mathrm{SG}}(p)\,|_{\substack{\text{single}\\ \text{state}}} = G^{x_1;x_1}_{1;1} - 2\,G^{x_1;x_1}_{1;x_1} + G^{x_1;x_1}_{x_1;x_1} = \frac{1}{p^2}, \tag{8.70}$$

$$\chi^{\mathrm{NL}}(p)\,|_{\substack{\text{single}\\ \text{state}}} = G^{x_1;x_1}_{1;1} - 4\,G^{x_1;x_1}_{1;x_1} + 3G^{x_1;x_1}_{x_1;x_1} = \frac{1}{p^2 + 2wQ_1}. \tag{8.71}$$

It turns out that the massless behaviour of the single state spin glass correlation Eq. (8.70) also holds when considering the exact replicon propagators, explicitly showing the occurrence of a Goldstone mode. More about it later (see Section 9.2.4).

8.7 Summary

In this chapter we have introduced a field theory for the spin glass in finite dimension with a field ϕ_i^{ab}. This field is governed by the 'truncated' Lagrangian, Eq. (8.5), where besides the marginal cubic w coupling one keeps a dangerous irrelevant quartic u coupling. Separating ϕ_i^{ab} into a mean field order parameter Q^{ab} and a fluctuation φ_i^{ab}, we have then investigated the effect of fluctuations and correlations thereof above $D = 6$, the upper critical dimension. Our main results can be summarized in the following statements.

- The RS mean field solution is unstable towards Gaussian fluctuations below the AT line, which results from the presence of a small negative mass for the replicon mode.

- The RS bare propagators in the replicon and longitudinal–anomalous sectors are easily obtained using the Replica Fourier Transform, that (quasi) diagonalizes the Hessian matrix.
- In the spin glass phase, where the RSB mean field solution has to be considered, the diagonalization of the Hessian and the inversion problem to obtain the bare propagators are much more difficult to perform. The main reason is the complex nature of the mean field solution and, consequently, of the Hessian matrix itself. This requires an appropriate parametrization of the Hessian and of the propagators to take account of the tree (ultrametric) structure of the RSB solution.
- The unitarity equations for the bare propagators in the RSB case are a set of complicated coupled integral equations. However, in the so-called near infrared regime, they simplify drastically and can easily be solved. The propagators exhibit two poles in this approximation, that eventually become two cuts in the exact solution. This regime gives an accurate description of the spin glass behaviour for dimensions $D > 6$.
- In the near infrared regime it is possible to compute the one-loop correction to the equation of state. The behaviour of the renormalized u coupling turns out to become singular below $D = 8$. As a consequence, mean field theory is found to be perturbatively stable only above $D = 8$, while for $6 < D < 8$ it has to be corrected to take account of the effects of the irrelevant coupling. Below $D = 6$ infrared divergences take over and the near infrared approximation is no longer admissible.
- The propagators have a direct interpretation in terms of correlation overlaps between pure states. Besides, the standard correlation functions can be written as sums of correlation overlaps and, therefore, of propagators.

References

Bray A. J. and Moore M. A. (1979). *J. Phys. C*, **12**, 79.
De Dominicis C. and Kondor I. (1983). *Phys. Rev. B*, **27**, 606.
De Dominicis C., Kondor I. and Temesvari T. (1998). Beyond the SK model. In *Spin Glasses and Random Fields*, ed. A. P. Young, Singapore, World Scientific.
Edwards S. F. and Anderson P. W. (1975). *J. Phys. F*, **5**, 965.
Fisher D. and Sompolinsky H. (1985). *Phys. Rev. Lett.*, **54**, 1063.
Harris A., Lubensky T. and Chen J. (1976). *Phys. Rev. Lett.*, **36**, 415.
Parisi G. (1983). *Phys. Rev. Lett.*, **50**, 1946.
Pytte E. and Rudnick J. (1979). *Phys. Rev. B*, **19**, 3603.
Temesvari T., Kondor I. and De Dominicis C. (1988). *J. Phys. A*, **21**, L1145.

9
Propagators, mostly replicon

In the previous chapter we have started to look at the difficult task of inverting the Gaussian fluctuation matrix around a Parisi-like RSB equilibrium Q^{ab}. As we have seen, the core of the difficulty is manifest in the unitarity equations (8.27). In Section 8.4 we have shown that in the near infrared regime these equations become linearly coupled equations that can easily be solved to get the propagators. In general, however, the unitarity equations contain nontrivial integrals over replica overlaps (see Eq. (8.37)). To overcome this problem and solve the equations in full generality one resorts once again to the technique of the Replica Fourier Transform. This is what we shall do in this chapter. Even if the complete resolution of the unitarity equations is beyond the scope of this book, we wish at least to compute explicitly the propagators in the most dangerous replicon sector, giving a few hints on what can be done for the other sectors. With this aim in mind, we proceed in the following way. In Section 9.1 we return to the RFT technique which was used in Chapter 6 to diagonalize the Hessian (in the absence of RSB) and we generalize it for R steps of RSB. We exhibit how it works on a simple toy model and in Section 9.2 we use it to obtain the propagators in the replicon sector. In Section 9.3 we use scaling arguments to guess what becomes of the most singular replicon propagators below and away from the upper critical dimension $D = 6$, obtaining thereby a lower critical dimension for the Parisi spin glass. In Section 9.4 we sketch the derivation of the propagators in the other sectors (longitudinal–anomalous), by further extending the RFT. Finally, in Section 9.5 we discuss the behaviour of the generalized susceptibility.

9.1 RFT diagonalization on a tree

In Chapter 6 we have introduced and applied the RFT method to diagonalize a quadratic form for a Hessian matrix with an RS structure. What we must do now is to generalize this technique to deal with more complicated RSB fluctuation matrices, which are the ones appearing at low temperature. The main complication lies in the fact that now the co-distances (or overlaps) among replicas, with respect to which the RFT is performed, have a nontrivial ultrametric structure and lie on a tree. The RFT must then be generalized to properly take into account this structure.

9.1.1 Coordinates on a tree

As a first step, it is convenient to identify a given replica index a through its coordinates on the tree. We then define

$$a : [a_0\, a_1 \;\cdots\; a_R] \tag{9.1}$$

as the coordinates of a. These coordinates tell the path to reach replica a on the tree, namely (see Fig. 6.3):

- from the root, follow the branch a_0 among the p_0/p_1 possible branches;
- then follow the a_1 branch among the following p_1/p_2 branches at level 1;
- ...
- follow the a_R branch among the $p_R/p_{R+1} \equiv p_R$ branches at level R.

Within this notation, two replicas a and b happen to have mutual overlap $a \cap b = r$ if they follow the same path until level r and then depart, i.e. if $a_0 = b_0, \ldots, a_{r-1} = b_{r-1}$, but $a_r \neq b_r$, that is

$$\begin{array}{r|l|}
a : & a_0\, a_1 \ldots a_{r-1}\; a_r \ldots a_R \\
 & \qquad\qquad\quad \neq \\
b : & a_0\, a_1 \ldots a_{r-1}\; b_r \ldots b_R
\end{array} . \tag{9.2}$$

9.1.2 Toy model diagonalization

We now use this notation to diagonalize via RFT a simple toy model for the Gaussian fluctuations. Consider the quadratic form

$$-\mathcal{H} = \frac{1}{2} \sum_{a \neq b} \varphi_a M^{ab} \varphi_b, \tag{9.3}$$

where M is a matrix with R steps of Replica Symmetry Breaking (such as, for example, the overlap matrix Q^{ab}). M^{ab} is only depending upon the (co)distance, i.e. the overlap between the two replicas a and b. This, contrary to the RS case studied so far, can take $R + 1$ possible values, so that

$$\begin{aligned} M^{ab} &= M_r \qquad \text{if } a \cap b = r; \\ M_{R+1} &= 0. \end{aligned} \tag{9.4}$$

Given the R-step structure of the matrix M, it is convenient to specify the replica index of the fields φ_a through its tree coordinates, as defined in Eq. (9.1). The fields then depend upon $R+1$ discrete variables. Each one of them, say $[a_j]$, takes p_j/p_{j+1} values, i.e. $(0, 1, 2, \ldots, p_j/p_{j+1}-1)$ on the circle, i.e. mod p_j/p_{j+1}. We can now generalize the RFT introduced in Chapter 6. To RFT with respect to replica a on a tree, we RFT each one of the $[a_j]$ variables specifying the coordinates of a on the tree. Thus, focusing on $[a_j]$,

$$\begin{cases} \varphi[a_j] = \dfrac{1}{\sqrt{p_j/p_{j+1}}} \displaystyle\sum_{\hat{a}_j} \mathrm{e}^{\frac{2\pi\mathrm{i}}{p_j/p_{j+1}} a_j \hat{a}_j} \varphi[\hat{a}_j], \\ \varphi[\hat{a}_j] = \dfrac{1}{\sqrt{p_j/p_{j+1}}} \displaystyle\sum_{a_j} \mathrm{e}^{-\frac{2\pi\mathrm{i}}{p_j/p_{j+1}} a_j \hat{a}_j} \varphi[a_j], \end{cases} \tag{9.5}$$

remembering that

$$\sum_{a_j} \mathrm{e}^{\frac{2\pi\mathrm{i}}{p_j/p_{j+1}} a_j \hat{a}_j} = \frac{p_j}{p_{j+1}} \delta_{\hat{a}_j;\hat{0}}. \tag{9.6}$$

The fields φ depend, however, on the whole string $[a_0 \ldots a_R]$ and we must write this dependence explicitly in order to RFT over all the a_i. To lighten the notation, from now on the symbol φ is omitted and the fields are specified directly through the corresponding strings. For clarity the string of $[a_j]$ standing for φ_a will be written upon the string $[b_j]$ associated with φ_b.

We then obtain, successively,

$$\begin{aligned} -\mathcal{H} &= \frac{1}{2} \sum_{r=0}^{R} M_r \sum_{\{a,b\}} \begin{matrix} [a_0 & \cdots & a_{r-1} & a_r & a_{r+1} & \cdots & a_R] \\ [a_0 & \cdots & a_{r-1} & b_r & b_{r+1} & \cdots & b_R] \end{matrix} \left(1 - \delta_{a_r;b_r}\right) \\ &= \frac{1}{2} \sum_{r=0}^{R+1} (M_r - M_{r-1}) \sum_{\{a,b\}} \begin{matrix} [a_0 & \cdots & a_{r-1} & a_r & a_{r+1} & \cdots & a_R] \\ [a_0 & \cdots & a_{r-1} & b_r & b_{r+1} & \cdots & b_R] \end{matrix} \end{aligned} \tag{9.7}$$

with $M_{-1} = 0$. Performing RFT, we get

$$\begin{aligned} -\mathcal{H} = \frac{1}{2} \sum_{r=0}^{R+1} \Bigg\{ (M_r - M_{r-1}) \sum_{\{\hat{a}\}} \begin{bmatrix} \hat{a}_0 \cdots & \hat{a}_{r-1} & \hat{0}_r & \cdots \hat{0}_R \\ -\hat{a}_0 \cdots & -\hat{a}_{r-1} & \hat{0}_r & \cdots \hat{0}_R \end{bmatrix} \\ \times \frac{p_r}{p_{r+1}} \frac{p_{r+1}}{p_{r+2}} \cdots \frac{p_R}{p_{R+1}} \Bigg\}. \end{aligned} \tag{9.8}$$

Each factor p_j/p_{j+1} results from

$$(p_j/p_{j+1})^2 \times \left(\frac{1}{\sqrt{p_j/p_{j+1}}} \right)^2, \tag{9.9}$$

where, for each $j \geq r$, the first factor comes from free summations (see Eq. (9.6)) and the second from normalizations (see Eq. (9.5)). With $p_{R+1} \equiv 1$, we then find

$$-\mathcal{H} = \frac{1}{2}\sum_{r=0}^{R+1} p_r\,(M_r - M_{r-1}) \sum_{\{\hat{a}\}} \begin{bmatrix} \hat{a}_0 & \cdots & \hat{a}_{r-1} & \hat{0}_r & \cdots \hat{0}_R \\ -\hat{a}_0 & \cdots - & \hat{a}_{r-1} & \hat{0}_r & \cdots \hat{0}_R \end{bmatrix}. \quad (9.10)$$

Contrary to what may seem, the fields $[\hat{a}_0 \ldots \hat{a}_{r-1}\,\hat{0}_r \ldots \hat{0}_R]$ are not orthonormal: indeed $\hat{a}_{r-1}$ is allowed the value $\hat{0}_{r-1}$, and in this case the string mixes with $[\hat{a}_0 \ldots \hat{a}_{r-2}\,\hat{0}_{r-1} \ldots \hat{0}_R]$, and hence with all of them. To get orthogonalized fields we still have to separate in the sum $\sum_{\hat{a}}$ between $\hat{0}$ components and $\hat{a}'(\hat{a}' \neq \hat{0})$ components. That is, in the sums over $\hat{a}_j$, we systematically separate the null value and the others,

$$\sum_{\hat{a}_j} \begin{bmatrix} \hat{a}_j \\ -\hat{a}_j \end{bmatrix} = \begin{bmatrix} \hat{0}_j \\ \hat{0}_j \end{bmatrix} + \sum_{\hat{a}'_j} \begin{bmatrix} \hat{a}'_j \\ -\hat{a}'_j \end{bmatrix}. \quad (9.11)$$

Consider now the components

$$\sum_{\{\hat{a}\}} \begin{bmatrix} \hat{a}_0 & \cdots & \hat{a}'_{r-1} & \hat{0}_r & \cdots & \hat{0}_R \\ -\hat{a}_0 & \cdots & -\hat{a}'_{r-1} & \hat{0}_r & \cdots & \hat{0}_R \end{bmatrix}, \quad (9.12)$$

and ask what is the $\sum p_t\,(M_t - M_{t-1})$ multiplying it? To answer that, we must list all the parent 'states' which under successive application of operation (9.11) would result in (9.12). Clearly the answer is

$$-\mathcal{H} = \frac{1}{2}\sum_{r=0}^{R+1} \left[\sum_{t=r}^{R+1} p_t\,(M_t - M_{t-1})\right] \sum_{\{\hat{a}\}} \begin{bmatrix} \hat{a}_0 & \cdots & \hat{a}'_{r-1} & \hat{0}_r \cdots \hat{0}_R \\ -\hat{a}_0 & \cdots & -\hat{a}'_{r-1} & \hat{0}_r \cdots \hat{0}_R \end{bmatrix}, \quad (9.13)$$

now in a truly diagonalized form with orthogonal fields (no more mixing). By definition then the coefficient can be called the Replica Fourier Transform on the tree of the matrix M_{ab} (and in fact it can be shown to be related to the standard Fourier Transform over a group, see Carlucci and De Dominicis, 1997):

$$\hat{M}_r \equiv \sum_{t=r}^{R+1} p_t (M_t - M_{t-1}). \quad (9.14)$$

The converse comes out as

$$M_r = \sum_{t=0}^{r} \frac{1}{p_t} (\hat{M}_t - \hat{M}_{t+1}), \quad (9.15)$$

with M and $\hat{M}$ satisfying

$$\hat{M}_r - \hat{M}_{r+1} = p_r (M_r - M_{r-1}). \quad (9.16)$$

The toy model in its diagonalized form is thus (upon reintroducing the field φ in the notation)

$$-\mathcal{H} = \frac{1}{2}\sum_{r=0}^{R+1} \hat{M}_r \sum_{\{\hat{a}\}} \left|\varphi\left[\hat{a}_0\ \hat{a}_1\ \ldots\ \hat{a}'_{r-1}\ \hat{0}_r\ \ldots\ \hat{0}_R\right]\right|^2 \tag{9.17}$$

with the field components $\varphi[\hat{a}_0\, \hat{a}_1 \ldots \hat{a}'_{r-1}\hat{0}_r \ldots \hat{0}_R]$ being orthogonalized.

In the continuum limit $R \to \infty$, $p_r \to p\,(x) = x$ in Parisi gauge, $M_r \to M\,(x)$ with $M_{R+1} \equiv 0$, $M_R \to M\,(x_1)$, and $p_{R+1} \equiv 1$, $p_0 \equiv n \to 0$, one recovers a transform introduced by Mézard and Parisi (1991). One has

$$\hat{M}(x) = \int_x^1 t\frac{\mathrm{d}M(t)}{\mathrm{d}t}\mathrm{d}t - M(x_1),$$
$$M(x) = \int_0^x \frac{1}{t}\frac{\mathrm{d}\hat{M}(t)}{\mathrm{d}t}\mathrm{d}t. \tag{9.18}$$

If the role of the matrix M is played, for example, by the overlap matrix $Q(x)$, then one has to consider that in the presence of a magnetic field $Q\,(x) = Q\,(x_0)$ for $x < x_0$, so that (9.18) becomes

$$Q\,(x) = Q\,(x_0) + \int_{x_0}^x \frac{1}{t}\frac{\partial\hat{Q}\,(t)}{\partial t}\mathrm{d}t. \tag{9.19}$$

9.1.3 Properties of the RFT on a tree

The RFT on the tree has the general properties common to ordinary Fourier Transform. For example, for what concerns convolutions, given $\sum_c M^{ac}M^{cb} \equiv (M^2)^{ab}$, we have

$$\widehat{(M^2)}_k = \hat{M}_k^2,$$
$$\widehat{(M^m)}_k = \left(\hat{M}_k\right)^m,$$
$$\widehat{(M^{-1})}_k = 1/\hat{M}_k. \tag{9.20}$$

In particular,

$$\operatorname{tr} M^m = n\sum_{k=0}^{R+1}\left(\frac{1}{p_k} - \frac{1}{p_{k-1}}\right)\left(\hat{M}_k\right)^m, \tag{9.21}$$

while, we recall,

$$\sum_{a,b}(M^{ab})^m = n\sum_{r=0}^{R+1}(p_r - p_{r+1})\,(M_r)^m, \tag{9.22}$$

where one notices the multiplicities

$$\mu(k) \equiv n\left(\frac{1}{p_k} - \frac{1}{p_{k-1}}\right) \to -n\frac{\mathrm{d}k}{k^2},$$
$$n\,\delta r \equiv n\,(p_r - p_{r+1}) \to -n\,\mathrm{d}x. \tag{9.23}$$

9.1.4 Equation of state via RFT

Let us now see how the RFT can be used to solve the equation of state. We use the properties analyzed immediately above on the overlap matrix Q^{ab}. The equation of state reads

$$-h^2 = 2\tau\, Q_r + w(Q^2)_r + \frac{2u}{3}Q_r^3, \tag{9.24}$$

where we have used $\sum_c Q^{ac}Q^{cb} = (Q^2)^{ab}$. We also have, using (9.16), and $(\widehat{Q^2})_k = (\hat{Q}_k)^2$,

$$(Q^2)_r = \sum_{k=0}^{r}\frac{1}{p_k}\left(\hat{Q}_k^2 - \hat{Q}_{k+1}^2\right) = \sum_{k=0}^{r}(Q_k - Q_{k-1})(\hat{Q}_k + \hat{Q}_{k+1}), \tag{9.25}$$

hence

$$-h^2 = 2\tau Q_r + w\sum_{k=0}^{r}(Q_k - Q_{k-1})\left(\hat{Q}_k + \hat{Q}_{k+1}\right) + \frac{2u}{3}Q_r^3. \tag{9.26}$$

Subtracting the same equation for $r-1$, we get

$$0 = (Q_r - Q_{r-1})\left\{2\tau + w\left(\hat{Q}_r + \hat{Q}_{r+1}\right) + \frac{2u}{3}\left(Q_r^2 + Q_r\,Q_{r-1} + Q_{r-1}^2\right)\right\}. \tag{9.27}$$

Note that if $r = 0$, in the continuum limit and in zero magnetic field, we get

$$\tau = -w\hat{Q}_0 \Longrightarrow +w\int_0^1 \mathrm{d}x\, Q(x). \tag{9.28}$$

Subtracting again from (9.27) the same equation for $r-1$, after division by $Q_r - Q_{r-1}$, we have

$$0 = w\left(\hat{Q}_{r+1} - \hat{Q}_{r-1}\right) + \frac{2u}{3}(Q_r - Q_{r-2})(Q_r + Q_{r-2} + Q_{r-1}). \tag{9.29}$$

Finally, we get the exact relationships

$$\hat{Q}_{r+1} - \hat{Q}_{r-1} = -p_{r-1}(Q_{r-1} - Q_{r-2}) - p_r(Q_r - Q_{r-1}), \tag{9.30}$$

where we have used (9.14) to compute the first term of (9.29), and

$$w[p_{r-1}(Q_{r-1} - Q_{r-2}) + p_r(Q_r - Q_{r-1})]$$
$$= \frac{2u}{3}(Q_r - Q_{r-2})\,[Q_r + Q_{r-2} + Q_{r-1}]\,, \tag{9.31}$$

which can be solved *exactly* using p_r stationarity (De Dominicis and Di Francesco, 2003). Here we are content to get

$$\left(1 + O\left(\frac{1}{R}\right)\right)\frac{wp_r}{2u} = Q_r, \tag{9.32}$$

and in the $R \to \infty$ limit, the Parisi result,

$$Q(x) = \frac{wp(x)}{2u}. \tag{9.33}$$

9.2 The propagators in the replicon sector

Let us now apply the technique developed in the previous sections to address our main problem. Given the Hessian matrix $M^{ab;cd}$ (8.26) appearing in the Lagrangian (8.25) we want to find its inverse, i.e. the propagators of the SG field theory (i.e. the fluctuation correlations). In this respect, as we shall see, the RFT is a powerful tool that allows us in the simplest way to find both the M eigenvalues and G, the inverse of M (De Dominicis, Carlucci and Temesvari, 1997). This is not the way the propagators were originally obtained, but it is the one involving least labour and most apt to generalization. The replicon propagators were also obtained by Parisi and Sourlas (2000) using p-adic algebra.

9.2.1 Parametrization of the replicon sector

We have already described in Section 8.3.1 how to parametrize our RSB matrices M (or G). Here we focus upon the most singular sector, the *replicon sector*, which will turn out to display Goldstone modes. Let us now summarize once again how to characterize that sector. From its definition we have:

$$M^{ab;cd} \equiv M^{r;r}_{u;v}, \tag{9.34}$$

where

$$\begin{aligned} &a \cap b = c \cap d = r, && r = 0, 1, \ldots, R, \\ &u = \max\,(a \cap c, a \cap d)\,, && u = r+1, r+2, \ldots, R,\ \ R+1, \\ &v = \max\,(b \cap c, b \cap d)\,, && v = r+1, r+2, \ldots, R,\ \ R+1. \end{aligned} \tag{9.35}$$

In terms of tree structure this corresponds to Fig. 9.1.

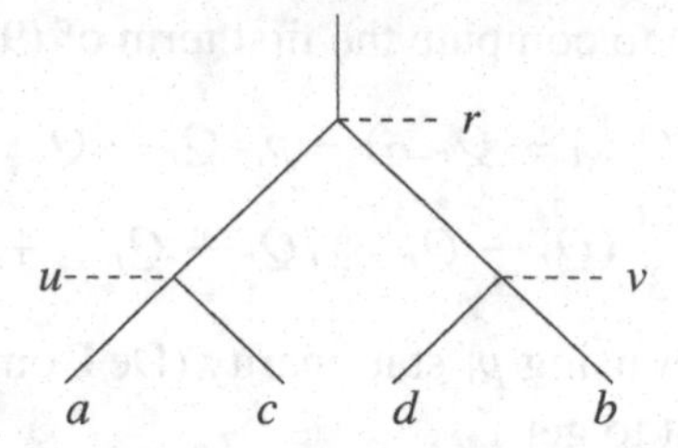

Figure 9.1 Tree structure of the replicon propagator

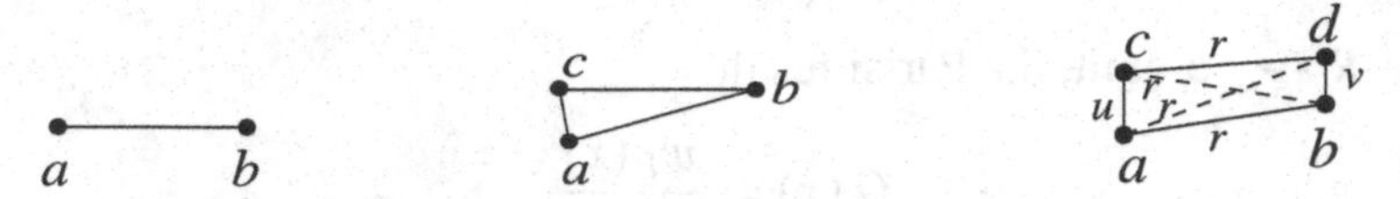

Figure 9.2 Ultrametric geometry for two replicas (left), three replicas (centre), four replicas in the replicon sector (right)

In terms of ultrametric geometry this sector also has a specific characterization. As we have already noted, given ultrametricity, all the triangles built with three replicas are isosceles. Imagine that we have two replicas with overlap (co-distance) $a \cap b = r$ (see Fig. 9.2, left). If we generate a third replica close to a (i.e. $a \cap c$ close to 1), we find that $(c \cap b) \equiv (a \cap b) \leq (a \cap c)$. In distances this corresponds to Fig. 9.2 (centre). Let us now add a fourth replica d. If d is close to *a and c* we have a geometry characterizing the *longitudinal–anomalous* sector. If it is close to b we have a case where $c \cap d = c \cap b = a \cap d = a \cap b = r$ and $a \cap c = u, d \cap b = v$, characterizing the *replicon sector* (see Fig. 9.2, right).

9.2.2 The solution via RFT

Our purpose is solve, for G, the unitarity equation

$$\frac{1}{2}\sum_{c;d} M^{ab;cd}\, G^{cd;ef} = \delta_{(ab);(ef)} \tag{9.36}$$

for G and M belonging to the replicon sector, and where M is therefore known as $M^{r;\,r}_{u;\,v}$. The index r plays here a passive role and it is necessary to perform the RFT *with respect to the cross-overlaps* (i.e. lower indices) *only*.

To understand how to proceed let us first compare our case with the one of a RFT on a *single index* field. The unitarity equation (9.36) is replaced in this case with

$$\sum_{c} m^{a;c} g^{c;e} = \delta_{a;e}, \tag{9.37}$$

where $m^{a;b} = m_r$ for $a \cap b = r$. Under RFT of both sides we get

$$\hat{m}_k \hat{g}_k = 1, \tag{9.38}$$

solved by

$$\hat{g}_k = 1/\hat{m}_k \tag{9.39}$$

with

$$\hat{m}_k = \sum_{t=k}^{R+1} p_t \left(m_t - m_{t-1}\right). \tag{9.40}$$

Finally, the wanted $g_r (g^{a;b} = g_r$ for $a \cap b = r)$ is obtained via an inverse RFT:

$$g_r = \sum_{k=0}^{r} \frac{1}{p_k} \left(\frac{1}{\hat{m}_k} - \frac{1}{\hat{m}_{k+1}} \right), \tag{9.41}$$

which solves the problem giving $g^{a;b} = g_r$ vs. $m^{a;b} = m_r$.

We have solved the problem for the matrices $m^{a;b}$, $g^{a;b}$ because we have noticed a convolution. If we wish to graphically mark the convolution between distances, we may write

$$\overline{a \quad ; c} \qquad \overline{c \; ; \quad e}$$

where we connected with a line the distances with respect to which the convolution holds. This is aptly reduced to a product via RFT, just as it happens with ordinary distances and FT. In Eq. (9.36) we have just a double convolution, i.e. graphically

$$\begin{array}{cc} \overline{a \quad ; c} & \overline{c \quad ; e} \\ \underline{b \; ; \quad d} & \underline{d \; ; \quad f} \end{array}$$

(or the alternative one by exchange of c, d). The *double* convolution can be now reduced to a product by a *double RFT*. Instead of (9.38) and (9.39) we have

$${}_{\mathrm{R}}G^{r;r}_{\hat{k};\hat{l}} = 1/\left({}_{\mathrm{R}}M^{r;r}_{\hat{k};\hat{l}}\right) \tag{9.42}$$

with what replaces (9.40),

$${}_{\mathrm{R}}M^{r;r}_{\hat{k};\hat{l}} = 1/\left({}_{\mathrm{R}}G^{r;r}_{\hat{k};\hat{l}}\right) = \sum_{u=k}^{R+1}\sum_{v=l}^{R+1} p_u p_v \left(M^{r;r}_{u;v} - M^{\;\;r;r}_{u-1;v} - M^{r;r}_{u;v-1} + M^{\;\;r;r}_{u-1;v-1}\right), \tag{9.43}$$

where we have used the sub-index R whenever the replicon component of G and M is considered. Conversely, knowing the double RFT (9.43) one can get back the replicon propagator (now replacing (9.41))

$$ {}_{\mathrm{R}}G^{r;r}_{u;v} = \sum_{k=r+1}^{k=u} \sum_{l=r+1}^{l=v} \frac{1}{p_k} \frac{1}{p_l} \left[{}_{\mathrm{R}}G^{r;r}_{\hat{k};\hat{l}} - {}_{\mathrm{R}}G^{r;r}_{\widehat{k+1};\hat{l}} - {}_{\mathrm{R}}G^{r;r}_{\hat{k};\widehat{l+1}} + {}_{\mathrm{R}}G^{r;r}_{\widehat{k+1};\, \widehat{l+1}} \right] \tag{9.44} $$

with ${}_{\mathrm{R}}G^{r;r}_{\hat{k};\hat{l}}$ as given in (9.42).

We note that in Eq. (9.43), defining the replicon RFT component, we have used the *full M* component which should be understood as

$$ M^{r;r}_{u;v} = {}_{\mathrm{R}}M^{r;r}_{u;v} + M^{r;r}_{u} + M^{r;r}_{v} - M^{r;r}_{r}. \tag{9.45} $$

Here the replicon component ${}_{\mathrm{R}}M^{r;r}_{u;v}$ exists only for $u, v \geq r+1$ and the longitudinal–anomalous component is given for all u by

$$ M^{r;r}_{u} = \lim_{s \to r} {}_{\mathrm{A}}M^{r;s}_{u}, \tag{9.46} $$

where the precise definition of ${}_{\mathrm{A}}M^{r;s}_{u}$ and its RFT is detailed below (Section 9.4). The only thing that matters here is that, being now a component with a *single cross-overlap*, it is projected out under a *double RFT*. Thus Eq. (9.43) is valid either with the full M,

$$ M = {}_{\mathrm{R}}M + {}_{\mathrm{A}}M, \tag{9.47} $$

or with its replicon counterpart ${}_{\mathrm{R}}M$ alone.

9.2.3 The replicon propagators in RFT

Let us now use the components of $M^{ab;\,cd}$ as they come out from the truncated model for $M^{ab;cd}$ (Section 8.3), that is

$$ \begin{aligned} M^{ab;ab} &= p^2 - 2(\tau + u(Q^{ab})^2), \\ M^{ab;ac} &= -wQ^{bc}, \quad b \neq c, \\ M^{ab;cd} &= 0, \qquad\quad (ab) \neq (cd). \end{aligned} \tag{9.48} $$

Using our parametrization $M^{r;r}_{u;v}$, $u;v \geqslant r+1$, this gives

$$ M^{\ \ r;r}_{R+1;R+1} = p^2 - 2\left(\tau + uQ_r^2\right), \tag{9.49} $$

$$ M^{\ \ r;r}_{R+1;\max(ab;bc)} = -wQ^{bc} = -wQ(v) = M^{\ \ r;r}_{R+1;v}, \tag{9.50} $$

$$ M^{\ \ r;r}_{\max(ac;ab);R+1} = -wQ^{ac} = -wQ(u) = M^{r;r}_{u;R+1}, \tag{9.51} $$

all other components (with $a \cap b = c \cap d \equiv r$) vanishing.

Let us now compute explicitly the double RFT in Eq. (9.43). The component (9.49) gives a contribution

$$ p^2_{R+1} M^{\ \ r;r}_{R+1;R+1} = M^{\ \ r;r}_{R+1;R+1} = p^2 - 2\left(\tau + uQ_r^2\right) \tag{9.52} $$

to the sum. The component (9.50) gives instead

$$p_{R+1}\sum_{u=k}^{R+1} p_u\left(M^{r;r}_{u;R+1} - M^{\;r;r}_{u-1;R+1}\right) = -w\sum_{u=k}^{R+1} p_u\left(Q_u - Q_{u-1}\right).$$
$$= -w\,\hat{Q}_k. \qquad (9.53)$$

Likewise for (9.51), so that altogether we get

$$M^{r;r}_{\hat{k};\hat{l}} = p^2 - w\big(\hat{Q}_k + \hat{Q}_l\big) - 2\big(\tau + uQ_r^2\big). \qquad (9.54)$$

Recalling, from the equation of state, the relationship $\tau = -w\hat{Q}_0$, this expression can be cast in the following more useful form†

$$M^{r;r}_{\hat{k};\hat{l}} = p^2 + w\big[\big(\hat{Q}_0 - \hat{Q}_k\big) + \big(\hat{Q}_0 - \hat{Q}_l\big)\big] - 2uQ_r^2,$$
$$k, l \geq r+1. \qquad (9.55)$$

At this point it is perhaps worth noticing that the *lowest* replicon eigenvalue $M^{r;r}_{\widehat{r+1};\,\widehat{r+1}}$ becomes *exactly* (De Dominicis and Di Francesco, 2003)

$$M^{r;r}_{\widehat{r+1};\,\widehat{r+1}} = \frac{u}{3}\left(\frac{Q_R - Q_0}{R}\right)^2, \qquad (9.56)$$

turning into a zero mode as $R \to \infty$. Let us now consider the continuum limit, $R \to \infty$. The order parameter and its RFT become two continuous functions: if we use $r \to x$, and keep using k, l for their continuum limit (see Eq. (9.18)) then $Q_r \to Q(x)$ and $\hat{Q}_k \to \hat{Q}(k)$. In the Parisi gauge $p_r \to p(x) = x$, Eq. (9.55) becomes for $x < k, l \leq x_1$,

$$w\hat{Q}(k) = w\int_k^{1-} x\frac{\mathrm{d}Q}{\mathrm{d}x}\mathrm{d}x - wQ(x_1) = w\int_k^{x_1} x\frac{\mathrm{d}Q}{\mathrm{d}x}\mathrm{d}x - wQ(x_1)$$
$$= -\frac{w^2}{4u}k^2 - wQ_1\left(1 - \frac{x_1}{2}\right), \qquad (9.57)$$

and thus

$$w\big(\hat{Q}(k) - \hat{Q}(0)\big) = -w\int_0^k x\frac{\mathrm{d}Q}{\mathrm{d}x}\mathrm{d}x = -\frac{w^2}{4u}k^2, \qquad (9.58)$$

which, together with

$$uQ^2(x) = +\frac{w^2}{4u}x^2, \qquad (9.59)$$

† Another way, perhaps simpler, is to subtract from (9.54) the equation of state.

gives for the double RFT

$$1/G^{x;x}_{\hat{k};\hat{l}} = M^{x;x}_{\hat{k};\hat{l}} = p^2 + \frac{w^2}{4u}(k^2 + l^2 - 2x^2),$$
$$x \le k, l \le x_1. \tag{9.60}$$

For k, l or $x > x_1$, the values of $Q(x)$ and $\hat{Q}(k)$ remain respectively frozen at $Q(x_1) = Q_1$ and, from (9.57), $\hat{Q}(x_1) = -Q_1$. In this case we have, instead of (9.58),

$$w\big(\hat{Q}(k) - \hat{Q}(0)\big) = \frac{w^2}{4u}x_1^2, \qquad k > x_1, \tag{9.61}$$

showing that in (9.60), when the variables k, l exceed the value x_1, they remain frozen at that value.

Equation (9.60) provides in explicit form the Hessian and its inverse (the propagators) under the double RFT. Note that in obtaining $M^{x;x}_{\hat{k};\hat{l}}$ we have obtained at the same time the eigenvalues of $M^{ab;cd}$ in the (most dangerous) replicon sector. Recall that for the RS solution ($R = 0$) we had, for the (single) replicon eigenvalue,

$$M_{\mathrm{R}} = p^2 - 2(\tau - wQ + uQ^2), \tag{9.62}$$

i.e. a *negative* eigenvalue which, using the equation of state, would be written

$$M_{\mathrm{R}} = p^2 - \frac{4}{3}uQ^2 + \frac{h^2}{Q}. \tag{9.63}$$

Here, using the full RSB solution, we have a spectrum of masses that extends to the interval

$$[0, \quad 2uQ(x_1)^2] \tag{9.64}$$

and, in the presence of a magnetic field (i.e. with the lower bound 0 becoming $x_0 > 0$), to the interval

$$[0, \quad 2u(Q(x_1)^2 - Q(x_0)^2)]. \tag{9.65}$$

In both cases the spectrum, of amplitude τ^2, is nonnegative, indicating the stability of the Parisi ansatz. At the same time, it exhibits the existence of *zero modes* (Goldstone modes) for values of $k = l = x$ or $x > x_1$.

9.2.4 The standard replicon propagators

The connection of propagators to correlation overlaps, which provides some sort of physical intuition, is simple in their original, replica representation $G^{ab;cd}(p)$. Thus, we may want now to return to it, performing the inverse (double) RFT. (For calculational purposes however the above RFTed representation is much simpler and it may be expedient working out loop corrections to stay in that representation.)

Then in this section we wish to compute

$$ {}_{\mathrm{R}}G^{r;r}_{u;v} = \sum_{k=r+1}^{k=u} \sum_{l=r+1}^{l=v} \frac{1}{p_k} \frac{1}{p_l} \left[G^{r;r}_{\hat{k};\hat{l}} - G^{r;r}_{\hat{k};\hat{l}} - G^{r;r}_{\widehat{\hat{k};l+1}} + G^{r;r}_{\widehat{k+1};\widehat{l+1}} \right]. \tag{9.66} $$

Going to the continuous limit, with $r \to x$, $u \to z_1$ and $v \to z_2$, we get for $x \leq z_1, z_2 \leq x_1$,

$$ {}_{\mathrm{R}}G^{x;x}_{z_1;z_2} = \int_x^{z_1} \frac{dk}{k} \frac{\partial}{\partial k} \int_x^{z_2} \frac{dl}{l} \frac{\partial}{\partial_l} \frac{1}{p^2 + \frac{w^2}{4u}[k^2 + l^2 - 2x^2]}. \tag{9.67} $$

Note that ${}_{\mathrm{R}}G^{x;x}_{\hat{k};\hat{l}}$ is continuous, but ${}_{\mathrm{R}}G^{x;x}_{z_1;z_2}$ is *not*. Indeed, from (9.66) one can see that there is a *jump* between the values computed $u = R$ (becoming $z_1 = x_1$ in the continuous limit) and $u = R + 1$ (becoming $z_1 = 1$). Speaking directly in the continuous limit, what happens is that the integrand in Eq. (9.67) assumes a constant value in the interval $[x_1; 1]$. In this way we find, for example,

$$ \begin{aligned} {}_{\mathrm{R}}G^{x;x}_{1;1} = & \int_x^{x_1} \frac{dk}{k} \frac{\partial}{\partial k} \int_x^{x_1} \frac{dl}{l} \frac{\partial}{\partial l} \frac{1}{p^2 + \frac{w^2}{4u}(k^2 + l^2 - 2x^2)} \\ & - 2 \int_x^{x_1} \frac{dk}{k} \frac{\partial}{\partial k} \frac{1}{p^2 + \frac{w^2}{4u}\left(k^2 + x_1^2 - 2x^2\right)} + \frac{1}{p^2 + \frac{w^2}{2u}\left(x_1^2 - x^2\right)}. \end{aligned} \tag{9.68} $$

Something analogous happens when only one of the two lower indices is fixed at 1.

There are two propagators that are worth mentioning. Let us consider first propagators inside a given state. These are obtained by setting $r = \mathrm{R}$, i.e. $x = x_1$ in the continuum. They describe fluctuations inside a given ergodic component. Using the previous formulas we find exactly

$$ {}_{\mathrm{R}}G^{x_1;x_1}_{1;1} = \frac{1}{p^2}, \tag{9.69} $$

that is a Goldstone mode, indicating that equilibrium states are actually marginal.

Other interesting propagators to look at are those with zero overlap, i.e. with $r = 0$ ($x = 0$ for a continuous RSB). They are given by

$$ \begin{aligned} {}_{\mathrm{R}}G^{0;0}_{x_1;x_1} &= \int_0^{x_1} \frac{dk}{k} \frac{\partial}{\partial k} \int_0^{x_1} \frac{dl}{l} \frac{\partial}{\partial l} \frac{1}{p^2 + \frac{w^2}{4u}(k^2 + l^2)} \\ &= 2 \left(\frac{w^2}{2u} \right)^2 \int_0^{x_1} \int_0^{x_1} \frac{dk\, dl}{\left(p^2 + \frac{w^2}{4u}(k^2 + l^2) \right)^3}. \end{aligned} \tag{9.70} $$

Letting

$$k = \sqrt{\frac{4u}{w^2}}\, p\, x, \qquad l = \sqrt{\frac{4u}{w^2}}\, p\, y, \tag{9.71}$$

we get

$$\begin{aligned} {}_{\mathrm{R}}G^{0;0}_{x_1;x_1} &= \frac{2w^2}{up^4} \int_0^{c/p} \int_0^{c/p} \frac{\mathrm{d}x\, \mathrm{d}y}{(1+x^2+y^2)^3} \\ &\underset{p\to 0}{\sim} \frac{w^2}{up^4} \int_0^{\pi/4} \mathrm{d}\theta \int_0^{\infty} \mathrm{d}\rho\, \frac{\mathrm{d}\rho^2}{(1+\rho^2)^3} \\ &= \frac{\pi}{4} \frac{w\, Q_1}{x_1} \frac{1}{p^4}. \end{aligned} \tag{9.72}$$

Using now the renormalized equation of state (see Section 8.5) we find

$$ {}_{\mathrm{R}}G^{0;0}_{x_1;x_1} \simeq \frac{w^2}{\tilde{u}\, p^4} \sim \frac{(\tau)^{(8-D)/2}}{w^2\, p^4}. \tag{9.73}$$

Note that the singular infrared behaviour is mainly due to the singular multiplicity near $k, l \sim 0$,

$$\mu\,(k) = \frac{1}{p_k} - \frac{1}{p_{k-1}} = \frac{p_{k-1} - p_k}{p_k\, p_{k-1}} \to -\frac{\mathrm{d}k}{k^2}. \tag{9.74}$$

The strong infrared behaviour of (9.73) was first noticed by Sompolinsky and Zippelius (1983). At the time it was considered to preclude the existence of a spin glass phase below $D = 4$. However, the behaviour exhibited in (9.73) is valid above $D = 6$. At $D = 6$ scaling is restored with

$$ {}_{\mathrm{R}}G^{0;0}_{x_1;x_1} \sim \frac{1}{p^2} \left(\frac{\tau}{p^2} \right), \tag{9.75}$$

and below one has to incorporate loop corrections. This is the strongest infrared behaviour exhibited by the replicon propagators, other components behaving at most as $1/p^3$ in the $p \to 0$ limit (see De Dominicis, Kondor and Temesvari, 1998; see also Ferrero and Parisi, 1996).

9.3 A hint at the lower critical dimension

In the previous section we have been able to compute explicitly the Gaussian propagators in the replicon sector. The effect of the interactions on the propagators is to generate corrections and the appearance of nontrivial critical exponents. The computation of loop corrections, however, is painful, and we content ourselves with watching the effect of introducing some simple scaling ansatz for the propagators.

Consider first what happens in the very simple case of the pure Ising system. At zero loop, in the ferromagnetic phase, the fluctuation matrix reads

$$M(p) = p^2 + u\frac{m^2}{2}, \tag{9.76}$$

where u is now the quartic coupling, m the magnetization ($m^2 \sim \tau^\beta$ with loop corrections). Under loop corrections we get for the leading terms

$$M(p) \simeq p^{2-\eta} + \left(u\frac{m^2}{2}\right)^{\gamma/2\beta} + \cdots = p^{2-\eta}\left[1 + \left(\tau/p^{1/\nu}\right)^\gamma + \cdots\right], \tag{9.77}$$

which leads to the Kadanoff scaling law,

$$G(p) \sim \frac{1}{p^{2-\eta}} f\left(\frac{\tau}{p^{1/\nu}}\right), \tag{9.78}$$

with

$$\begin{aligned} f(x) &\sim \frac{1}{x^\gamma} && \text{for } x \to \infty \quad (p \to 0), \\ f(0) &\sim C && \text{for } \tau = 0. \end{aligned} \tag{9.79}$$

Likewise, we now consider the most singular replicon propagator exhibiting the correlation overlap between far apart states (i.e. of zero mutual overlap), see Eq. (9.73). The corresponding RFT fluctuation matrix is

$$M^{0,0}_{\hat{k},\hat{l}}(p) = p^2 + \frac{wQ_1}{2x_1}(k^2 + l^2), \tag{9.80}$$

for which we want to know how loop corrections modify this zero loop contribution.

As $k = l = 0$ we have a Goldstone mode, i.e. massless. This mode is usually understood as having a p^2 behaviour, but the Goldstone theorem only states that the mass is vanishing (in the absence of the conjugate field). This p^{-2} behaviour is generally true for models with a continuous symmetry. The reason is that generally the η exponent is *positive*, hence powers in $p^{2-\eta}$ are less infrared divergent. In the *spin glass case η is negative*, and the Goldstone modes' behaviour will be different. To put it another way, what happens in $O(N)$ systems is that the effective coupling for the transverse modes (after integrating out the massive modes) is vanishing with p^2, meaning that transverse (massless) modes are free in the infrared (see Chapter 10). Here the Goldstone modes are the *bottom of a band* (of width of the small mass), and they remain coupled to the massive, yet transverse, modes of that band, and hence the Goldstone mode is not infrared free. As a result, loop corrections will change the p^2 behaviour and we tentatively write

$$M^{0,0}_{\hat{k},\hat{l}}(p) \simeq p^{2-\eta} + \left[\frac{wQ_1}{x_1}(k^2 + l^2)\right]^{\gamma/\beta} + \cdots \tag{9.81}$$

to obtain the scaling in $1/p^{2-\eta}$ or in $1/\tau^{\gamma}$ for $G^{0;0}_{\hat{k},\hat{l}}(p)$. If we integrate now the ansatz (9.81) over the multiplicity of k, l, as in (9.67) with $z_1 = z_2 = x_1$ and $x = 0$, we get

$$
{}_{\mathrm{R}}G^{0;0}_{x_1;x_1} \underset{p\to 0}{\sim} \frac{\pi}{4}\frac{w Q_1}{x_1}\frac{1}{p^{\frac{D+2-\eta}{2}}}\frac{s}{2s-1},
$$

$$
s = \frac{4-2\eta}{D-2+\eta}, \tag{9.82}
$$

which extends (9.73) below $D = 6$. On that ansatz the lower critical dimension D_{lc} is obtained when (9.82) diverges like $1/p^D$, and thus

$$
D_{\mathrm{lc}} = 2 - \eta(D_{\mathrm{lc}}). \tag{9.83}
$$

With the value of η obtained from numerical estimates (Kawashima and Young, 1996; Katzgraber and Young, unpublished) one comes close to the value

$$
D_{\mathrm{lc}} = \frac{5}{2} \tag{9.84}
$$

obtained by Franz, Parisi and Virasoro (1994) from a self-consistent mean field calculation and confirmed recently (Franz and Toninelli, 2004, 2005). Of course, one has now to check, via loop calculations of $M^{0;0}_{\hat{k};\hat{l}}(p)$, that the ansatz (9.81) is reasonable. The calculation, even at one loop, is quite complicated (Carlucci, 1997).

9.4 What about the other sectors?

Using the RFT method, we have been able to invert the Hessian matrix and compute the propagator in the replicon sector. For the longitudinal–anomalous (L–A) sector the job is more difficult. First of all, it turns out that the best we can do is to block-diagonalize the Hessian into blocks of size $R \times R$, while the ultimate task is to invert a matrix of much larger size $n(n-1)/2 \times n(n-1)/2$. But even to do that, the RFT that worked so well in the replicon sector is not enough. It has to be generalized to operate on the tree when more replicas than the pair directly concerned by the Fourier Transform are involved. The four replicas occurring in the replicon sector were actually operating independently two by two (see Fig. 9.3): in $M^{ab;cd} \equiv M^{r;r}_{u;v}$ the cross-overlaps $u, v > r$ are equivalent and carry the same degeneracy, so that a double RFT can be performed with respect to them without ambiguity. As we shall see, this is not the case in the L–A sector.

Here we content ourselves by characterizing the L–A sector and hinting at the result of the inversion, for more details the reader is referred to the review of De Dominicis *et al.* (1998).

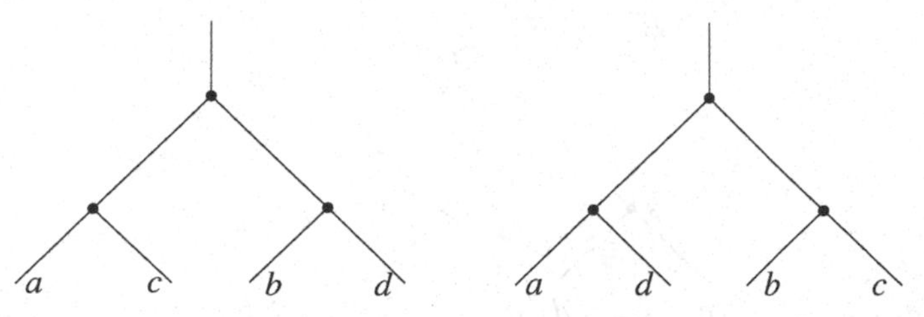

Figure 9.3 Tree representation for the replicon sector. The two figures show the two possible structures compatible with the replicon geometry

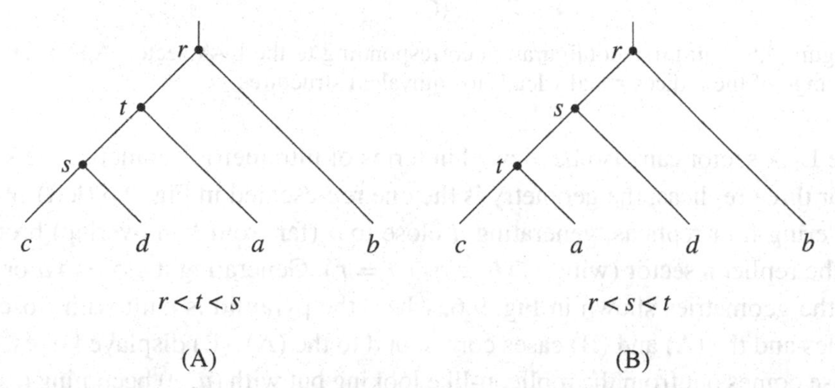

Figure 9.4 The first two possible tree configurations corresponding to the longitudinal–anomalous sector. Exchanging r and s leads to equivalent structures

We start by recalling the parametrization for the L–A sector introduced in the previous chapter:

$$M^{ab;cd} = M_t^{r;s}, \tag{9.85}$$

with

$$\begin{aligned} r &= a \cap b, \\ s &= c \cap d, \\ t &= \max(a \cap c,\ a \cap d,\ b \cap c,\ b \cap d). \end{aligned} \tag{9.86}$$

The L–A sector matrix is then characterized by the fact that it has a *single* lower (i.e. cross-overlap) index (while in the R sector the matrix has two lower indices and two identical upper (overlap) indices).

In terms of tree structure, just as we have drawn the single tree that gives rise to the replicon sector, in Figs. 9.4 and 9.5 we draw the three different tree configurations (A, B and C) that occur for the longitudinal–anomalous sector. Note that configuration C is only formally like one of the replicon sectors, since overlaps and cross-overlaps are exchanged here.

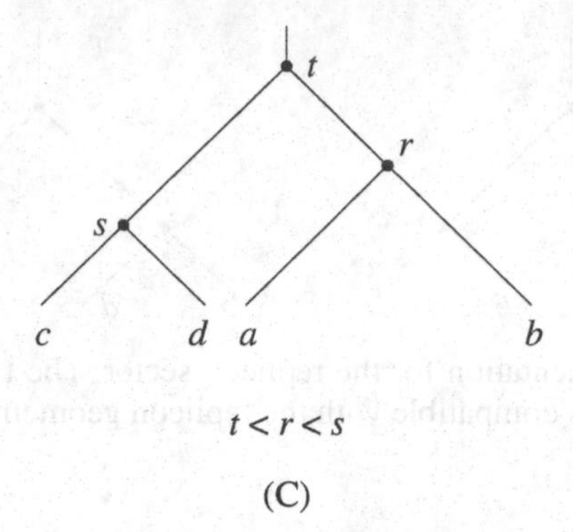

Figure 9.5 Third tree configuration corresponding to the L–A sector. Again, exchange of the indices r and s leads to equivalent structures

The L–A sector can also be viewed in terms of ultrametric geometry. We know that for three replicas, the geometry is the one represented in Fig. 9.6 (left). When considering four replicas, generating d close to b (far from b in overlap) brought us to the replicon sector (with $a \cap b \equiv c \cap d = r$). Generating it close to a or to c gives the geometries shown in Fig. 9.6, where the pyramid is built with isosceles triangles and the (A) and (B) cases correspond to the (A), (B) displayed trees. The (C) case comes out from the replicon-like looking but with (u, v) becoming (r, s) – see Fig. 9.2.

Now, if we want to block-diagonalize the Hessian matrix, again the appropriate RFT applied to the lower index t (the cross-overlap) will do the job. This time, however, we must take into account when summing over p_ts the presence of overlaps r and s, that is we must account for the branching off that occurs when t crosses the *fixed* (passive) values r and s. In the R sector there was no such crossing since $u, v \geq r+1$ in $M^{r;r}_{u;v}$. This can be done by defining the *extended* RFT (in the presence of r, s) by

$$M^{r;s}_{\hat{k}} = \sum_{t=k}^{R+1} p_t^{(r,s)} \left(M^{r;s}_t - M^{r;s}_{t-1}\right) \tag{9.87}$$

with, if e.g. $r < s$,

$$p_t^{(r,s)} = \begin{array}{l} p_t, \;\; t \leq r, \\ 2p_t, \; r < t \leq s, \\ 4p_t, \; r < s < t. \end{array} \tag{9.88}$$

The inverse transform is defined as, with the same rules,

$$M^{r;s}_t = \sum_{k=0}^{k=t} \frac{1}{p_k^{(r,s)}} \left(M^{r;s}_{\hat{k}} - M^{r;s}_{\widehat{k+1}}\right). \tag{9.89}$$

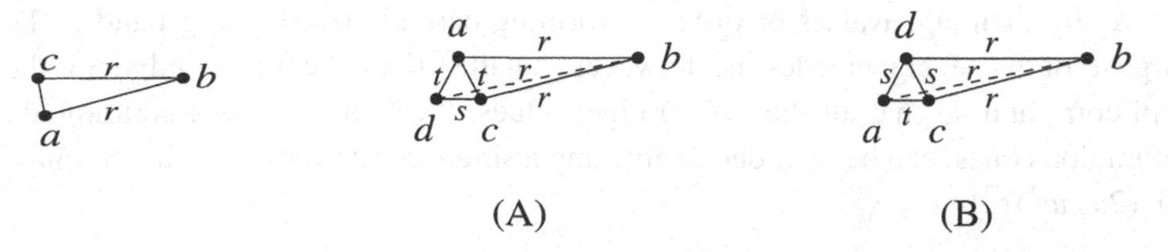

Figure 9.6 Geometry in the L–A sector: three replicas (left) and four replicas (right, cases A and B)

In the special case we are interested in (the truncated model) one finds

$$M^{ab;ac} = M^{r;s}_{R+1} = -wQ(\min(r,s)). \tag{9.90}$$

The relationship relating $M^{r;s}_{\hat{k}}$ to $G^{r;s}_{\hat{k}}$ (the Dyson equation) is easily derived with the RFT to give

$$G^{r;s}_{\hat{k}} = -g_{\hat{k}}(r)\left[M^{r;s}_{\hat{k}}\, g_{\hat{k}}(s) + \sum_{t=0}^{R} M^{r;t}_{\hat{k}} \frac{\delta_t^{(k-1)}}{4} G^{t;s}_{\hat{k}}\right], \tag{9.91}$$

where

$$g_{\hat{k}}(r) = \begin{cases} G_{\underbrace{r+1;\,r+1}}^{\;r;r} & \text{if } k \le r+1, \\ G_{\underbrace{r+1;\hat{k}}}^{\;r;r} & \text{if } k > r+1, \end{cases} \tag{9.92}$$

and with the notation

$$\delta_t^{(k-1)} = p_t^{(k-1)} - p_{t+1}^{(k-1)}. \tag{9.93}$$

The integral equation above can be solved exactly (via the use of Gegenbauer functions) whenever $M^{r;s}_{\hat{k}}$ is only depending upon k and $\min(r, s)$ or $\max(r, s)$. A fully explicit form for the solution $G^{r;s}_{\hat{k}}$ can be found in De Dominicis *et al.* (1998). From $G^{r;s}_{\hat{k}}$ one then obtains $G^{r;s}_t$ via the inverse RFT, thereby completing the set of spin-glass propagators. Note that, in the absence of a magnetic field, $M^{r;s}_{R+1}$ is vanishing with $\min(r, s)$ and so is $G^{r;s}_{\hat{k}}$. Note also that $g_{\hat{k}}(r)$ is $1/p^2$ for $k \le r+1$. As a result the (replica) components ${}_{\mathrm{A}}G^{x;y}_t$ are found to contain $1/p^3$ infrared divergences, but nothing is known on loop corrections, i.e. on what this $1/p^3$ divergence becomes below $D = 6$.

In the same way it is possible to obtain exact expressions for the eigenvalues for the L–A sector. One finds a first set of eigenvalues which are of order τ and form a continuous band in the interval

$$\left[2\tau\left(1 - \frac{u}{3w^2} + \cdots\right),\ 2\tau\left(1 + \frac{u}{3w^2} + \cdots\right)\right]. \tag{9.94}$$

There are then eigenvalues of order τ^2 forming partially overlapping bands. The largest of these eigenmodes is, however, smaller than the upper edge of the replicon band so that all the $O(\tau^2)$ eigenvalues, the replicon and longitudinal–anomalous ones, can be regarded as forming a single continuous band in the range $[0, (2u/w^2)\tau^2]$.

9.5 The generalized susceptibility

To conclude this chapter we want to show how the complex structure of the propagators and their mass spectrum is reflected in physically measurable quantities. Once again, we focus on correlations and susceptibilities.

For convenience, we define a generalized correlation function, or generalized susceptibility, as

$$\begin{aligned}\chi^{\alpha}(r) &= C_1(r) - 2\,\alpha C_2(r) + (2\alpha - 1)\,C_3(r)\\ &= \left\{\overline{\langle S_i S_j\rangle^2} - 2\alpha\overline{\langle S_i S_j\rangle\langle S_i\rangle\langle S_j\rangle} + (2\alpha - 1)\overline{\langle S_i\rangle^2\langle S_j\rangle^2}\right\},\end{aligned} \tag{9.95}$$

where $r = r_i - r_j$. The brackets indicate, as usual, thermodynamical averages, here in the presence of an infinitesimal magnetic field. For $\alpha = 1$ we recover the spin glass correlation function $\chi^{\mathrm{SG}}(r)$ and for $\alpha = 2$ the nonlinear correlation function $\chi^{\mathrm{NL}}(r)$ (see Section 8.6.2, Eqs. (8.62) and (8.62)).

Using the expressions for the connected correlation functions obtained in Section 8.6, we get (in momentum space)

$$\begin{aligned}\chi^{\alpha}(p) &= \delta^{Kr}_{p;0}\ \chi^{\alpha}_{\mathrm{mf}} + \chi^{\alpha}_{\mathrm{fl}}(p)\\ &= \frac{2-\alpha}{3n(n-1)}\left[\sum_{ab} Q_{ab}^2 + \sum_{abc} Q_{ab}Q_{ac}\right]\delta^{Kr}_{p;0}\\ &\quad + \frac{1}{n(n-1)}\sum_{a,b}\left\{G^{ab;ab}(p) - \frac{2\alpha}{(n-2)}\sum_{c\neq b} G^{ab;ac}(p)\right.\\ &\quad\left. + \frac{(2\alpha-1)}{(n-2)(n-3)}\sum_{c,d\neq a,b} G^{ab;cd}(p)\right\}.\end{aligned} \tag{9.96}$$

The mean field part is easily computed, and gives

$$\begin{aligned}\chi^{\alpha}_{\mathrm{mf}} &= \frac{2-\alpha}{3}\left[\int_0^1 \mathrm{d}x\ Q^2(x) - \left(\int_0^1 \mathrm{d}x\ Q(x)\right)^2\right]\\ &= \frac{2-\alpha}{3}\frac{w^2}{4u^2}\left[\frac{x_1^3}{3} - \frac{x_1^4}{4}\right],\end{aligned} \tag{9.97}$$

which is the simplest consequence of lack of self-averaging of the $P_J(Q)$, i.e. $\overline{P_J(Q_1)P_J(Q_2)} \neq \overline{P_J(Q_1)}\ \ \overline{P_J(Q_2)}$.

Using our parametrization in sectors for an R-step RSB ansatz, $\chi_{\rm fl}^{\alpha}$ can be written as

$$\begin{aligned}
\chi_{\rm fl}^{\alpha} = -\frac{2-\alpha}{3}\Bigg\{&\sum_0^R \delta r \sum_0^R \delta s\ G^{rs}_{R+1} + \frac{1}{4}\sum_0^R \delta r\ G^{rr}_{\widehat{r+1}} \\
&+\sum_0^R \delta r\ {}_{\rm R}G^{rr}_{R+1,\widehat{r+1}} + \sum_0^R \delta r\ G^{rr}_{R+1,R+1}\Bigg\} \\
-\frac{2\alpha-1}{3}\Bigg\{&\sum_0^R \delta r\ G^{r,r}_{\widehat{r+1},\ \widehat{r+1}} + \frac{1}{2}\sum_0^R \delta r \sum_0^R \delta s\ G^{r,s}_{\hat{0}}\Bigg\},
\end{aligned} \tag{9.98}$$

where, we recall, $\delta r = p_r - p_{r+1}$ and $\delta s = p_s - p_{s+1}$.

From expressions (9.97) and (9.98) we note two general features which are peculiar to the RSB ansatz. First, contrary to the RS case (see Section 8.6.2) there is a finite mean field contribution. This means that the generalized correlation does *not* decay to zero (for $\alpha \neq 2$) at large distances and is a main consequence of ergodicity breaking. Second, we can clearly see from Eq. (9.98) that all the sectors contribute to the fluctuation term, and not only the replicon sector as was the case for the RS solution (see Eq. (8.69)).

Equation (9.98) can be explicitly computed using the expressions for the R and L–A propagators discussed in the previous sections. What is found is that (see De Dominicis *et al.*, 2005) a different behaviour holds for $\alpha = 2$ and $\alpha \neq 2$. More precisely,

$$\chi_{\rm fl}^{\alpha\neq 2} \sim \frac{A + \log p}{p^3} + \frac{B}{p^2}, \tag{9.99}$$

where A and B are numerical coefficients. The case $\alpha = 2$, on the other hand, exhibits a much less singular decay. This behaviour is already visible in the far infrared, where the $1/p^3$ power law becomes $1/p$, and is enhanced in the near infrared where the susceptibility displays a finite mass. Altogether, we have

$$\begin{aligned}
\chi_{\rm fl}^{\alpha=2}(p) &\simeq \frac{w x_1}{\sqrt{u}\,p}\left(\frac{p^2}{2x_1 w Q_1}\right)^{1/2} \sim \frac{1}{p}, && p^2 \ll 2x_1 w Q_1, \\
\chi_{\rm fl}^{\alpha=2}(p) &\simeq \frac{1}{p^2 + 2wQ_1(1-x_1)}, && 2x_1 w Q_1 \ll p^2 \ll 2wQ_1.
\end{aligned} \tag{9.100}$$

Thus, for intermediate momenta (or intermediate distances, i.e., between the correlation lengths associated with the two masses), $\chi_{\rm fl}^{\alpha=2}(p)$ behaves as if it were massive. Only at very large distance does one feel the infrared singularity in $1/p$ (or the associated inverse power behaviour in distance). This behaviour, predicted for $D \gtrsim 6$, might also hold in lower dimensions, and numerical simulations for $D = 3$ show for $\alpha = 2$ a fast exponential decay compatible with Eq. (9.100) (while the far infrared behaviour is not seen, suggesting a value for the large correlation length comparable to the system size).

For intra-state correlations we have already seen (see Eqs. (8.70) and (8.71)) that the analogues of $\chi^{\text{SG}} = \chi^{\alpha=1}$ and $\chi^{\text{NL}} = \chi^{\alpha=2}$ are respectively singular in $1/p^2$ and massive in $1/(p^2 + 2wQ_1)$. The complete correlation function (9.95) is an overlap average and both intra- and inter-state correlations contribute. As a result $\chi^{\alpha=1}$ has an enhanced singularity and $\chi^{\alpha=2}$ tends towards massiveness ($1/p^2$ is cancelled but we are left with a $1/p$ singularity and a massive-like behaviour in the near infrared).

9.6 Summary

In this chapter we have generalized the RFT technique to use it on a tree. In this way we have been able to diagonalize and invert the Hessian matrix in the replicon sector. We have found that

- the replicon propagators display a continuous spectrum of masses forming a band between zero and τ^2;
- the propagators inside a given state exhibit a Goldstone mode, revealing the marginal nature of the equilibrium states and fitting the scenario of a massless phase, with power-law decaying transverse correlations and massive longitudinal-like modes;
- zero overlap replicon propagators exhibit a $1/p^4$ infrared behaviour, the strongest infrared behaviour observed among all propagators.

The analysis of the longitudinal–anomalous sector is more subtle, and we only have sketched the first steps in this direction. The starting point is the extension of the RFT on a tree to the case where passive overlaps are present. We have not given details of the derivation of the propagators, but quoted results for propagators and masses (respectively inverse and eigenvalues of the Hessian matrix). Characteristically, the masses extend into two continuous bands (one around the large mass, the other contiguous to zero mass). Finally, in Section 9.5 we have come back to susceptibilities (the spin glass and the nonlinear susceptibilities) where one can watch contributions of intra-state and inter-state correlations.

References

Carlucci D. (1997). Ph.D. Thesis, Pisa.
Carlucci D. and De Dominicis C. (1997). *C. R. Acad. Sci. Paris*, **325**, 527.

De Dominicis C. and Di Francesco P. (2003). *J. Phys. A*, **36**, 10955.
De Dominicis C., Carlucci D. and Temesvari T. (1997). *J. Phys. I France*, **7**, 105.
De Dominicis C., Kondor I. and Temesvari T. (1998). Beyond the SK model. In *Spin Glasses and Random Fields*, ed. A. P. Young, Singapore, World Scientific.
De Dominicis, Giardina I., Marinari E., Martin O. and Zuliani F. (2005). *Phys. Rev. B*, **72**, 014443.
Ferrero M. E. and Parisi G. (1996). *J. Phys. A*, **29**, 3795.
Franz S. and Toninelli F. L. (2004). *J. Phys. A*, **37**, 7459.
Franz S. and Toninelli F. L. (2005). *J. Stat. Mech. Theor. Exp.*, P01008.
Franz S., Parisi G. and Virasoro M. A. (1994). *J. Phys. I France*, **4**, 1657.
Kawashima N. and Young A. P. (1996). *Phys. Rev. B*, **53**, R484.
Mézard M. and Parisi G. (1991). *J. Phys. I France*, **1**, 809.
Parisi G. and Sourlas N. (2000). *Eur. J. Phys.*, **14**, 535.
Sompolinsky H. and Zippelius A. (1983). *Phys. Rev. Lett.*, **50**, 1297.

10

Ward–Takahashi Identities and Goldstone modes

In the previous chapter we have explicitly computed components of the (bare) propagators in the replicon sector and found that we had a spectrum of masses bounded below by zero modes. In this chapter we show that these massless modes are indeed Goldstone modes related to the spontaneous breaking of a continuous symmetry of the Lagrangian, and thus are to remain massless under loop corrections. We have already analyzed a case of spontaneous symmetry breaking in the different context of the TAP free energy in Chapter 7. Here, we focus on the invariance properties of the spin glass Lagrangian and exhibit the continuous symmetry spontaneously broken below the AT line, which entails corresponding Ward–Takahashi Identities. We start in Section 10.1 by describing the problem in the simpler setting of a system with a two-component order parameter. In Section 10.2 we generalize to the spin glass case, by defining the symmetry which undergoes spontaneous breaking. In Section 10.4 we give a derivation of the Ward–Takahashi Identities, expressing the fact that the massless transverse modes are indeed Goldstone modes. Details of what follows can be found in De Dominicis, Kondor and Temesvari (1998).

10.1 The Legendre Transform and invariance properties

To explain the general problem we would like to address, it is convenient to consider first the case of a simpler field theory (at least for what concerns the notation). Let us work then with a field ϕ_i^a, its average value $Q_i^a = \langle \phi_i^a \rangle$ and a conjugate external source h_i^a. Let us consider, for example, a rotationally invariant system with n components (e.g. $a = 1, \dots, n$, where, in the simplest case, $n = 2$). If we call $F = -\ln Z$ the Helmholtz free energy, its Legendre Transform is defined by

$$\Gamma\left\{Q_i^a\right\} = F\left\{h_i^a\right\} + \sum_i \sum_a h_i^a Q_i^a, \tag{10.1}$$

and we have

$$Q_i^a = -\frac{\partial F}{\partial h_i^a} \quad \text{and} \quad \frac{\partial \Gamma}{\partial Q_i^a} = h_i^a. \tag{10.2}$$

Under an arbitrary rotation $\mathcal{R}$ ($n \times n$ orthogonal matrix) the Lagrangian and $\Gamma\{Q^a\}$ are invariant (in zero external source, or if the source undergoes the same rotation). So if

$$\vec{Q}' = \mathcal{R}\vec{Q}, \tag{10.3}$$

the invariance of Γ under $\mathcal{R}$ is written

$$\Gamma\{\vec{Q}'\} = \Gamma\{\vec{Q}\}. \tag{10.4}$$

We can write the invariance of Γ for its derivatives as

$$\frac{\partial \Gamma}{\partial Q^a}\{\mathcal{R}\vec{Q}\} = \sum_{a'} \mathcal{R}_{aa'} \frac{\partial \Gamma\{\vec{Q}\}}{\partial Q^{a'}}, \tag{10.5}$$

$$\frac{\partial^2 \Gamma\{\mathcal{R}\vec{Q}\}}{\partial Q^a \partial Q^b} = \sum_{a'b'} \mathcal{R}_{aa'}\mathcal{R}_{bb'} \frac{\partial^2 \Gamma\{\vec{Q}\}}{\partial Q^{a'} \partial Q^{b'}}, \tag{10.6}$$

etc. This can also be read as saying that on derivatives one can rotate either the internal indices (left hand side) or the external ones (right hand side) and get the same answer. These identities are always valid for any invariance of Γ. However, *if we have a continuous symmetry*, then *we can construct an infinitesimal* rotation and those identities become more interesting. Indeed, separating in $\mathcal{R}$ the identity

$$\mathcal{R} = \mathbb{I} + \delta\mathcal{R} \tag{10.7}$$

from (10.6), we get, expanding to lowest order in $\delta\mathcal{R}$,

$$\sum_b \frac{\partial^2 \Gamma}{\partial Q^a \partial Q^b} (\delta R\vec{Q})_b = (\delta R\vec{h})_a. \tag{10.8}$$

To understand better to what this leads, let us consider the simplest case with $n = 2$. The rotation matrix is in this case

$$\delta\mathcal{R} = \begin{pmatrix} 0 & 1 \\ -1 & 0 \end{pmatrix} \tag{10.9}$$

and $\vec{Q} = (Q_1, Q_2)$. Then $\delta R\vec{Q} = (Q_2, -Q_1)$ and (10.8) becomes

$$\begin{aligned} \Gamma_{11}Q_2 - \Gamma_{12}Q_1 &= h_2, \\ \Gamma_{21}Q_2 - \Gamma_{22}Q_1 &= -h_1, \end{aligned} \tag{10.10}$$

where the lower indices of Γ stand for derivatives with respect to components $a = 1$ and $a = 2$. Hence, for a magnetic field $\vec{h} = (h_1; 0)$ and thus a magnetization $(Q_1, 0)$, one finds:

$$\Gamma_{12} = \frac{\partial^2}{\partial Q_1 \partial Q_2} \Gamma\{\vec{Q}\}\big|_{Q_2=0} = 0, \tag{10.11}$$

$$\Gamma_{22} = \frac{h_1}{Q_1}. \tag{10.12}$$

These are the Ward–Takahashi Identities related to the symmetry (rotational in this case) of the Lagrangian. They tell us that we have a diagonal Hessian ($\Gamma_{12} = 0$) and that the *transverse propagator* has a *mass* h_1/Q_1, which in *zero* field, and below T_c when a spontaneous magnetization arises, *does vanish*. The spontaneous symmetry breaking of a *continuous* invariance group gives rise to a *massless* mode, the Goldstone mode.

10.2 The case of the spin glass

Formally, very little is to be changed in the case of the spin glass. We have a field ϕ_i^{ab}, its average value $Q_i^{ab} = \langle \phi_i^{ab} \rangle$ and its associated external source h_i^{ab} (in the absence of source, but with a magnetic field h, we have $h_i^{ab} = h^2$, i.e. replica independent). We also have an invariance of the Lagrangian and of $\Gamma\{Q^{ab}\}$ under the permutation group of our n replicas, P_n. The only question is how can we go beyond Eqs. (10.5) and (10.6) and define an infinitesimal permutation.

The first remark is to observe that we work with a set of $n\,(n-1)/2$ components, and, in the space of $n\,(n-1)/2$ dimensions, a permutation exchanges one axis with another, so we are still dealing with rotations. Thus,

$$Q' = PQ, \tag{10.13}$$

$$PQ^{ab} \equiv Q^{P_a P_b}. \tag{10.14}$$

If we split P such that

$$P = \mathbb{I} + \delta P, \tag{10.15}$$

$$\delta P Q^{ab} = Q^{P_a P_b} - Q^{ab}, \tag{10.16}$$

then (10.6) becomes

$$\frac{1}{2} \sum_{cd} \frac{\partial^2 \Gamma}{\partial Q^{ab} \partial Q^{cd}} \left(\delta P Q^{cd}\right) + \cdots = (\delta P h^{ab}). \tag{10.17}$$

If we are able to define a δP such that $\left(\delta P Q^{cd}\right)$ becomes infinitesimal, then one can drop all the other terms of the Taylor expansion and recover a standard Ward–Takahashi Identity.

10.3 Defining a small permutation

In the previous chapter we have parametrized each replica by its *address*

$$a : [a_0, a_1, \dots, a_{r-1}, a_r, \dots, a_R], \tag{10.18}$$

i.e. by the set of its $R+1$ coordinates, with

$$a_r = 1, 2, \dots, p_r/p_{r+1}. \tag{10.19}$$

Let us consider now permutations at *level r*, with $P^{(r)}$ defined by its action on the address

$$P^{(r)}a \equiv P^{(r)}[a_0, a_1, \dots, a_{r-1}, a_r, a_{r+1}, \dots, a_R], \tag{10.20}$$

and $P^{(r)}$ such that it acts in a nontrivial way *only if* a_{r+1} is *fixed* at a given value, say

$$a_{r+1} \equiv 1, \tag{10.21}$$

and *then it will change* a_r into $1 + a_r$, mod(p_r/p_{r+1}). In all other instances, $P^{(r)} \equiv \mathbb{I}$, i.e.

$$\begin{aligned} P^{(r)}&[a_0, a_1, \dots, a_{r-1}, a_r, a_{r+1}, \dots, a_R] \\ &\equiv \delta_{a_{r+1};1}[a_0, a_1, \dots, a_{r-1}, 1 + a_r, a_{r+1}, \dots, a_R] \\ &\quad + (1 - \delta_{a_{r+1};1})[a_0, a_1, \dots, a_{r-1}, a_r, a_{r+1}, \dots, a_R], \end{aligned} \tag{10.22}$$

as illustrated in Fig. 10.1. On the tree, as displayed in this figure, we can see that our choice amounts to a *circular permutation* $a_r \to 1 + a_r$, for *branches* a_r and their *single daughter branch* $a_{r+1} = 1$ and *all its respective descendants* (shaded on the figure).

With this definition, let us calculate the action of $\delta P^{(r)}$ on Q^{ab}, with

$$\delta P^{(r)} Q^{ab} = Q^{P_a^{(r)};P_b^{(r)}} - Q^{ab}. \tag{10.23}$$

Clearly, we have $\delta P_{ab}^{(r)} \neq 0$, only if in the address of a (or b) we have $a_{r+1} = 1$. If this is our choice, then we have two possibilities:

(i) $Q^{ab} = Q_{r+1}$, hence by definition $a_{r+1} \neq b_{r+1}$ (and $a_r = b_r$, $a_{r-1} = b_{r-1}, \dots$). Under $P^{(r)}$, $a_r \to 1 + a_r$ (for the address of a only, *not* for b), and hence $1 + a_r \neq b_r$, thus

$$\delta P^{(r)} Q^{ab} = Q_r - Q_{r+1}. \tag{10.24}$$

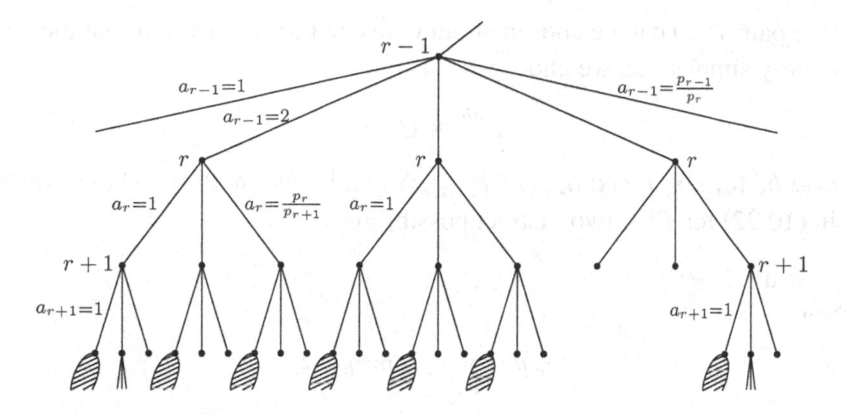

Figure 10.1 Tree representation of the permutation described in the text. Only the branches with $a_{r+1} = 1$ are permuted

(ii) $Q^{ab} = Q_r$, i.e. $a_r \neq b_r$ and $a_{r+1} = 1$. If $b_{r+1} = 1$ also, $\delta P^{(r)} Q^{ab} = 0$. If $b_{r+1} \neq 1$ one gets

$$\delta P^{(r)} Q^{ab} = Q_{r+1} - Q_r, \tag{10.25}$$

provided a_r *is such that* $1 + a_r = b_r$ (if not, (10.25) is *also* obtained by choosing the shift on a_r such that it brings it onto b_r).

In *all other cases* $P^{(r)} = \mathbb{I}$, i.e. $\delta P^{(r)} = 0$.

The net result is that we have with (10.22) constructed a permutation acting on the field Q^{ab} in such a way that its deviation from the identity is infinitesimal in the limit $R \to \infty$, since in that limit

$$Q_r - Q_{r+1} \sim O\,(1/R) \tag{10.26}$$

(except if $r = R$), thus allowing us to ignore all other terms in the Taylor expansion of (10.17). We have built classes of infinitesimal permutations that now generate for us Ward–Takahashi Identities. A general classification of infinitesimal permutations can be found in Goltsev (1991).

10.4 The simplest Ward–Takahashi Identity

From (10.17), when choosing a $P \equiv \mathbb{I} + \delta P$ with an infinitesimal δP as above, we have

$$\frac{1}{2} \sum_{cd} \frac{\partial^2 \Gamma}{\partial Q^{ab} \partial Q^{cd}} \left(\delta P^{(r)} Q^{cd}\right) = \left(\delta P^{(r)} h^{ab}\right) + O\left(\frac{1}{R^2}\right), \tag{10.27}$$

where the pair (a, b) can be chosen at our convenience. It turns out that the choice (i) is slightly simpler, i.e. we choose

$$Q^{ab} = Q_{r+1} \tag{10.28}$$

with $a_s \equiv b_s$ for $s \leqslant r$ and $a_{r+1} \neq b_{r+1}$. We still have, with the choice we have made in (10.22) for $P^{(r)}$, two distinct possibilities.

(i) a_{r+1} and $b_{r+1} \neq 1$
Then

$$\delta P^{(r)} Q^{ab} = \delta P^{(r)} h^{ab} = 0 \tag{10.29}$$

and the right hand side of (10.27) vanishes. Effecting the sum over all (c, d) pairs will give the simplest Ward–Takahashi Identity which is written (recall that the second derivative of Γ is nothing else than the Hessian matrix M introduced in the previous chapter)

$$\left[\frac{p_{r+2}}{4}\left(M^{r+1;r+1}_{\widehat{r+2}} - M^{r+1;r}_{\widehat{r+2}}\right) + O\left(\frac{1}{R^2}\right)\right](Q_r - Q_{r+1}) = 0. \tag{10.30}$$

This equation expresses that the RFT of the longitudinal–anomalous component (with a single lower index, i.e. cross-overlap) is continuous under variation of its upper indexes (overlaps).

(ii) $a_{r+1} = 1, \quad b_{r+1} \neq 1$ (or the converse)
Then

$$\delta P^{(r)} h^{ab} = h_r - h_{r+1}. \tag{10.31}$$

Effecting the sum over (c, d) pairs, one gets two types of contribution, both proportional to $Q_r - Q_{r+1}$. The first contribution is exactly equal to (10.30), which we can thus forget (since it is $O(1/R^2)$). The other contribution is in

$$M^{r+1;r+1}_{\widehat{r+2};\widehat{r+2}}(Q_r - Q_{r+1}) = (h_r - h_{r+1}) + O\left(\frac{1}{R^2}\right). \tag{10.32}$$

Here we recognize the replicon component under double RFT, i.e., Eq. (9.60) for the associated propagator (at zero momentum):

$$1/G^{\;r\,;\,r}_{\widehat{r+1};\,\widehat{r+1}} = M^{\;r\,;\,r}_{\widehat{r+1};\,\widehat{r+1}} = (h_r - h_{r-1})\,/(Q_r - Q_{r-1}), \tag{10.33}$$

which vanishes in zero source (i.e. for $h_r - h_{r-1} = 0$).

We already knew from direct calculation that the (truncated) Lagrangian was giving rise to a vanishing mass for the replicon propagator. Indeed, we found that in the $R \to \infty$ limit

$$M^{r;r}_{\hat{k};\hat{l}} \xrightarrow[R\to\infty]{} M^{x;x}_{\hat{k};\hat{l}}\,(p) = p^2 + \frac{w^2}{4u}[k^2 + l^2 - 2x^2]. \tag{10.34}$$

Hence, at zero momentum, and for $k = l = x$,

$$M^{x;x}_{\hat{x};\hat{x}}(0) = 0. \tag{10.35}$$

Here we have a proof that relies only upon invariance under permutations. It also tells us that in the presence of an external source h^{ab}, the Goldstone mode acquires a mass. Indeed, let us take the continuum limit of (10.33): r becomes $x = \lim_{R\to\infty} r/(R+1)$, while $r+1$ becomes $\lim_{R\to\infty}(r+1)/(R+1) = x + \lim_{R\to\infty} 1/(R+1) = x + 0$. We then get

$$M^{x;x}_{\hat{x};\hat{x}}(0) = \dot{h}(x)\,/\,\dot{Q}(x), \tag{10.36}$$

that only vanishes if h^{ab} is *overlap independent*, spelling that the Goldstone mode persists in the presence of a magnetic field, below the AT line.

10.5 A very special Goldstone mode

Standard Goldstone modes (as in Heisenberg or O(n) systems) have very simple properties in the infrared limit. Indeed, from Ward–Takahashi it then follows that transverse modes become *free* in that limit. One can check that property easily by asking for the effective transverse coupling in the infrared limit.

Let us consider, for example, the isotropic Heisenberg Model with n components (see, e.g., Brezin and Wallace, 1973). The field is in this case a vector $\vec{\phi}$ with n components. At the fluctuation level one finds in this case $n-1$ transverse (T) (Goldstone) modes and one longitudinal (L) mode, with propagators

$$G^{\mathrm{L}}(p) = \frac{1}{p^2 + \frac{g}{3}m^2},$$
$$G^{\mathrm{T}}(p) = \frac{1}{p^2}, \tag{10.37}$$

where $m = \langle \phi_{\mathrm{L}} \rangle$ is the magnetization, and g is the coupling constant associated with the quartic coupling:

$$\frac{g}{4!}\left(\vec{\phi}\cdot\vec{\phi}\right)^2 = {}^{\mathrm{T}}_{\mathrm{T}}\!\!>\!\bullet\cdots\bullet\!<\!\!{}^{\mathrm{T}}_{\mathrm{T}} + {}^{\mathrm{T}}_{\mathrm{T}}\!\!>\!\bullet\cdots\bullet\!<\!\!{}^{\mathrm{L}}_{\mathrm{L}} + {}^{\mathrm{L}}_{\mathrm{L}}\!\!>\!\bullet\cdots\bullet\!<\!\!{}^{\mathrm{L}}_{\mathrm{L}} \tag{10.38}$$

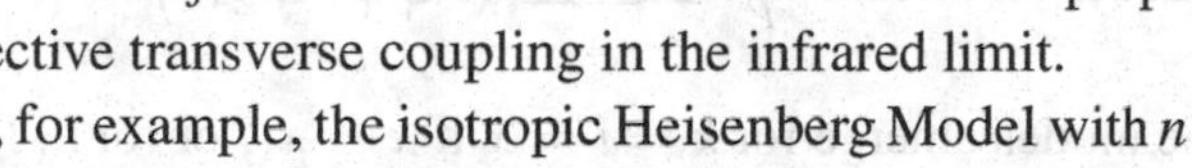

To investigate the behaviour of the transverse modes in the infrared, let us ask what is the effective interaction between the transverse modes, once the longitudinal mode has been integrated out. We then consider the four-point transverse vertex function $\Gamma^{(4)}_{\mathrm{TTTT}}$, and write the one-irreducible (*with respect to transverse lines*)

contributions to it

$$\Gamma^{(4)}_{\mathrm{TTTT}} = \begin{smallmatrix}\mathrm{T}\\ \\ \mathrm{T}\end{smallmatrix}\!\!>\!\!\bullet\cdots\bullet\!\!<\!\!\begin{smallmatrix}\mathrm{T}\\ \\ \mathrm{T}\end{smallmatrix} + \begin{smallmatrix}\mathrm{T}\\ \\ \mathrm{T}\end{smallmatrix}\!\!>\!\!\bullet\cdots \underset{\mathrm{L}}{\overset{\wr\quad\wr}{\text{———}}} \cdots\bullet\!\!<\!\!\begin{smallmatrix}\mathrm{T}\\ \\ \mathrm{T}\end{smallmatrix} + \cdots, \tag{10.39}$$

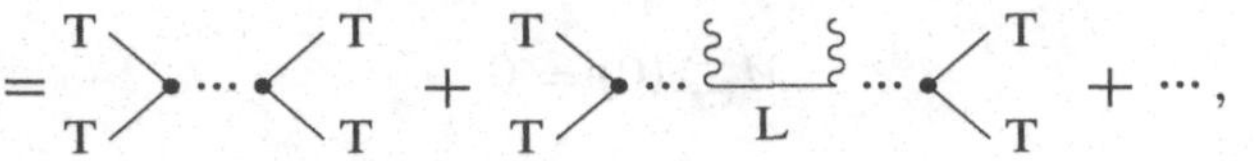

where the wavy lines stand for the magnetization and the internal L line for the bare longitudinal propagator. This reads as

$$\Gamma^{(4)}_{\mathrm{TTTT}} = g - \frac{1}{3}\frac{(gm)^2}{p^2 + \frac{g}{3}m^2} + \cdots = O(p^2). \tag{10.40}$$

Hence, in the infrared limit, not only do the Goldstone modes have zero mass, they also have a *vanishing coupling*. Their p^2 behaviour is unaffected by loop contributions at $T < T_{\mathrm{c}}$. At $T = T_{\mathrm{c}}$ *all modes* are massless, resulting in a $p^{2-\eta}$ behaviour.

What about the transverse modes in the spin glass case? As we have seen above, the Goldstone mode is, in this case, *not an isolated mode* but the bottom of a *band* characterized by three parameters (k, l and x) with a mass proportional to $k^2 + l^2 - 2x^2$ and a propagator (under double RFT)

$$G^{x;x}_{\hat{k};\hat{l}} = \frac{1}{p^2 + \dfrac{wQ_1}{x_1}(k^2 + l^2 - 2x^2)}. \tag{10.41}$$

The Goldstone modes emerge for $k = l = x$ if $x < x_1$, or for $k, l \geqslant x_1$ for $x = x_1$. As a result, it is no longer possible to write out an effective transverse mode Lagrangian resulting from integrating out the massive modes (there is no gap between the Goldstone modes and the rest of the band). Hence, in the infrared limit the Goldstone modes are *not free*, and thus, they do not behave as p^2, but as $p^{2-\eta}$. In zero field, if η is to depend only upon dimension and symmetries, it has to be the same as the one calculated at $T = T_{\mathrm{c}}$.

10.6 Summary

In this chapter we have exploited the invariance of the Lagrangian under the permutation group of n replicas to obtain some Ward–Takahashi Identities for the replicated effective field theory. A crucial step has been the definition of an appropriate permutation which becomes infinitesimal in the limit of infinite RSB.

Our main result is that the Goldstone mode acquires a mass only for overlap dependent sources $h(x)$, and thus persists in presence of a magnetic field, below the AT line. This Goldstone mode has special features distinct from those usually found in standard field theories. The reason is that it is *not* an isolated mode, but rather the bottom of a gapless band. As a consequence, massive modes cannot be integrated out as in O(n) models and the Goldstone modes remain coupled in the infrared.

References

Brézin E. and Wallace D. J. (1973). *Phys. Rev. B*, **7**, 986.
De Dominicis C., Kondor I. and Temesvari T. (1998). *J. Phys. IV, France*, **8**, 13. Preprint† cond-mat/9802166.
Goltsev A. V. (1991). *J. Phys. A*, **24**, 307.

† Equation numbering having been messed up at the editing stage, the reader should rather consult the cond-mat version.

11
Alternative approaches and conclusions

11.1 The droplet picture

Throughout this book we have developed a spin glass field theory for the fluctuation field around the mean field RSB Parisi solution. The justification for such a theory, as we have stated at the beginning of our discussion, is intimately related to the validity in finite dimension of the nontrivial multi-ergodic physical scenario depicted by the mean field solution. Whether this is the case or not is still not clear. We have commented in the chapters previous to this conclusion what are the main features of the spin glass field theory, the consistency it exhibits, some reasonable extrapolations to dimensions lower than six and some possible predictions for measurable observables. Despite the great effort and the rich results obtained so far, the validity of the theory in three dimensions is still, however, a debated point, as much as the assumption of a nontrivial spin glass state at low temperature. It therefore seems important to us to briefly mention a few alternative points of view where the low temperature phase has different features from the ones we have extensively described.

11.1.1 The droplet model

The richest and most interesting alternative picture for the EA model was developed along the years by various authors (Bray and Moore, 1984, 1986; McMillan, 1984; Fisher and Huse, 1986, 1987, 1988; Newman and Stein, 1992, 1996, 1998) and is generally referred to as the 'droplet model', from the Fisher and Huse paper of 1986. The physical scenario described by this model is in striking contrast to the mean field one. According to it, in the ordered phase at finite temperature only one pair of ergodic equilibrium states exists (related by time-reversal symmetry) instead of the number $O(N)$ predicted by the mean field picture. The difference between the two scenarios can be better understood by looking at the limit of infinitely many dimensions. In this limit, the results of the droplet picture are reproduced by the Migdal–Kadanoff approximation of the EA model (Gardner, 1984), and in this approximation the ground state can be found in linear time, contrary to the NP hardness appearing in the SK model. Still, the spin glass thermodynamic behaviour is nontrivial because of a nontrivial excitation spectrum.

The model assumes that the lowest energy excitations from the ground state around a particular point are compact clusters, or droplets, of order L^D spins coherently flipped, with the following properties (Fisher and Huse, 1986):

- droplets of size L with zero energy occur with probability $\sim 1/L^\theta$, where the stiffness exponent θ can be bounded between zero and $(D-1)/2$ (Fisher and Huse, 1988);
- the typical free energy of an excitation of size L scales as L^θ.

These features can be simply encoded by using an appropriate scaling ansatz for the probability distribution of the free energy F_L of size L droplets, namely

$$P_L(F_L) \sim \rho\left(\frac{F_L}{\mathcal{Y}L^\theta}\right)\frac{1}{L^\theta}, \tag{11.1}$$

with $\rho(0) > 0$, and where $\mathcal{Y}$ is a generalized stiffness modulus. Various numerical simulations have estimated the exponent θ for *compact* excitations: in $D = 3$ one has $\theta \sim 0.27$ (Carter, Bray and Moore, 2002) or $\theta \sim 0.24$ (Boettcher, 2004, 2005). In $D = 2$, on the other hand, it is negative (e.g. $\theta \sim -0.28$, Boettcher, 2004), implying the absence of ordered phase. The thermodynamical properties can be obtained with simple arguments by observing that, at finite temperature, the ordered phase is dominated by the thermally active droplets, i.e. droplets with $F_L \leq T$ and whose density is described by (11.1). For example, the average correlation functions can be computed by considering the probability that a pair of spins separated by distance r belong to an active droplet of linear dimension $L \geq r$.

11.1.2 Scaling theory and the behaviour of space correlations

All the results concerning the droplet model can be obtained (and extended) also within the framework of a scaling picture (Bray and Moore, 1986) and using a Renormalization Group terminology.

The concept of droplets of size L is in this case replaced by the one of 'block spin' on scale L, and instead of looking at the energy distribution of droplets of size L, one has to consider the distribution of the scale dependent effective couplings between block spins at scale L. For large length scales this distribution is supposed to approach a fixed universal shape, with a scale dependent width $J_L \sim L^\theta$. The distribution of $P_L(J_L)$ is given precisely by Eq. (11.1) (with J_L in place of F_L).

Using scaling arguments at $T = 0$ and at finite temperature one can compute many properties of the disordered phase (even critical ones) and recover the results of the droplet model. For example, let us consider the connected correlation functions $C_1(r)$, $C_2(r)$ and $C_3(r)$ introduced in Section 8.6.2 and defined as:

$$\begin{aligned} C_1(r) &= \overline{\langle S_i S_j \rangle^2} - \overline{\langle S_i \rangle^2}\ \overline{\langle S_j \rangle^2}, \\ C_2(r) &= \overline{\langle S_i S_j \rangle \langle S_i \rangle \langle S_j \rangle} - \overline{\langle S_i \rangle^2}\ \overline{\langle S_j \rangle^2}, \\ C_3(r) &= \overline{\langle S_i \rangle^2 \langle S_j \rangle^2} - \overline{\langle S_i \rangle^2}\ \overline{\langle S_j \rangle^2}, \end{aligned} \tag{11.2}$$

with $r_i - r_j = r$.

Within the scaling approach one has to consider block spins of size $L \geq r$ such that the sites i and j belong to the same block. In this way one can distinguish the contribution to correlations coming from *intra*-block fluctuations, and the one coming from block to block fluctuations. Clearly, this last one is determined by the distribution of the couplings at scale $L = r$, and in particular by the distribution of the internal field h_r acting on a block spin due to its neighbouring blocks (which has the same scaling form as the $P_L(J_L)$). At finite temperature only active blocks with $h_r < T$ will contribute. Integrating from zero to T, at large distances the leading contribution is due to $\rho(0)$ and one gets

$$\begin{aligned} C_1(r) \sim C_2(r) \sim C_3(r) &\sim Q_{\text{EA}}^2 \frac{T}{\mathcal{Y} r^\theta}, \\ C^{\text{SG}}(r) = C_1(r) - 2C_2(r) + C_3(r) &\sim Q_{\text{EA}}^2 \frac{T}{\mathcal{Y} r^\theta}, \end{aligned} \tag{11.3}$$

where we have also exhibited the behaviour for the spin glass correlation function.

Note that these correlation functions decay to zero at long distances. This is in contrast with the spin glass field theory where the large distance limit of correlation functions is nonvanishing (except for two special cases to which we return below) and given by the mean field prediction, as discussed in Section 9.5. A slightly different behaviour is found for the correlation function $C^{\text{NL}}(r) = C_1(r) - 4C_2(r) + 3C_3(r)$, that is for the space-dependent nonlinear susceptibility. In this case the leading term proportional to $\rho(0)$ vanishes and one has to consider sub-leading contributions. These are related to the shape of the scaled distribution $\rho(x)$ around $x = 0$. If, for small x, $\rho(x) \sim \rho(0) + Ax^\phi$, then one finds

$$C^{\text{NL}}(r) \sim Q_{\text{EA}}^2 \left(\frac{T}{\mathcal{Y} r^\theta} \right)^{1+\phi}. \tag{11.4}$$

These results could be compared with the prediction of the spin glass field theory which indeed allows for a large distance vanishing limit of the mean field-like correlation functions in two special cases:

(i) the spin glass correlation at zero overlap behaves as $1/r^2$ at $D = 6$ and, through scaling, becomes $1/r^{(D-2+\eta)/2}$ for $D < 6$ (see Sections 9.2.4 and 9.3);
(ii) the nonlinear correlation behaves as $1/r^5$ at $D = 6$ and as $1/r^{(5/4)(D-2+\eta)}$ for $D < 6$ (see Sections 9.5 and 9.3).

It is puzzling to realize that the numerical values of these exponents remain in the vicinity of what is obtained by numerical experiments, e.g., in $D = 3, \theta \sim 0.24$ and $(D-2+\eta)/2 \sim 0.3$. Besides, under a magnetic field these two special correlation functions no longer vanish at large distance. Whether these behaviours should be considered as a fluke or whether they carry some deep meaning remains fully unclear.

11.1.3 General predictions

Some major general predictions of the droplet picture are the following.

- In the ordered phase only one pair of symmetry related pure states exists, as signalled by the long distance vanishing limit of the correlation functions (Bray and Moore, 1986; Huse and Fisher, 1987). This corresponds to a trivial overlap distribution $P(Q)$ (i.e. a delta function in $\pm Q_{\mathrm{EA}}$).
- The presence of an external magnetic field suppresses the transition to the ordered phase and there is no AT line. To see this, one can generalize to the droplet model the Imry–Ma argument used in Chapter 2. The magnetization of a droplet of size L scales as $L^{D/2}$, thus the cost to flip the droplet is $L^\theta - hL^{D/2}$. Since $\theta < D/2$ we see that, according to this picture, the magnetic field always destroys the long range order and no AT line is predicted.
- The system responds in a chaotic way to (even small) changes in temperature. This is related to the highly nontrivial nature of the excitations, whose energy increases in a sub-additive way (being $\theta < D - 1$). To produce the L^θ scale dependence, different parts of the droplet boundary must give alternating contributions that altogether produce large cancellations, with a delicate balance between energy and entropy. Clearly, such balance is very sensitive to changes in temperature, giving rise to a chaotic T dependence (Fisher and Huse, 1988). We note that chaos in temperature is expected also within the mean field scenario, according to recent analytical results (Rizzo and Crisanti, 2003) and numerical simulations with large system sizes (Billoire and Marinari, 2002; Marinari, private communication, 2005).
- The droplet model, at least in the Migdal–Kadanoff limit, does not have aging in the response function (i.e. it corresponds to $X_{\mathrm{ag}} = 0$) (Ricci-Tersenghi and Ritort, 2000). This has to be compared with the mean field scenario where, as for the SK model, the

violation factor $X_{ag}(C)$ is nonzero and has a continuous dependence on the correlation function (Cugliandolo and Kurchan, 1994).

Of course many attempts have been made to distinguish between the mean field and the droplet scenario, using both analytical and numerical tools. Still, we cannot say the question has been settled and there is no agreement in the community. Numerical Monte Carlo simulations seem to favour the validity of the mean field scenario also for finite dimensional models (Marinari, Parisi and Ruiz-Lorenzo, 1998). However, criticisms have been put forward against these results, based on the Migdal–Kadanoff approximation, related to finite size and critical effects for the range of parameters investigated so far (Moore, Bokil and Drossel, 1998; Bokil *et al.*, 1999). Analytical thermodynamic arguments for the nonstability in finite dimension of an RSB structure have been constructed in support of the droplet picture (Newman and Stein, 1996, 1998, 2002), while analytical arguments based on the concept of stochastic stability (Franz *et al.*, 1999a,b) and the similarity between the static and dynamic behaviour have been invoked to support the mean field scenario.

11.1.4 Zero temperature excitations and the TNT picture

More recently, people have focused their attention upon the zero temperature behaviour and the detailed features of the excitation spectrum. Indeed, besides the thermodynamic predictions described above, the droplet model also has important implications on the geometric properties of the excitations.

The assumption of compact droplets implies that the excitation boundary scales as L^{D_s} where the fractal dimension D_s must satisfy the relation $D - 1 < D_s < D$. In other words the interface of the excitation has zero density in the thermodynamic limit.

It may be interesting to compare these features with the corresponding ones predicted by the mean field picture. If we assume that a multi-state scenario does hold, then there must exist excitations involving a finite fraction of the whole volume that have zero energy (indeed, by flipping $O(N)$ spins we can pass from a ground state to another one with the same free energy density). In the droplet language this means that there exist excitations with $\theta = 0$. These excitations are, however, *not* compact, in the sense that their boundary grows with an exponent $D_s = D$ and is therefore space-filling. This feature is related to the distribution of the so-called 'link-overlap' Q_l, which is defined as

$$Q_l = \frac{1}{N_b} \sum_{\langle ij \rangle} S_i S_j S_i' S_j', \qquad (11.5)$$

where primed variables are relative to the excited configuration and N_b stands for the number of bonds. For the SK model one trivially has $Q_l = Q^2$, and therefore the distribution $P(Q_l)$ is nontrivial. In terms of topology of the excitations, from the definition (11.5) we see that the link-overlap is simply one minus the fraction of links on the boundary of the excitation, which scales as $L^{D_s - D}$. This means that within the mean field scenario the excitations must have $D_s = D$ or, conversely, that a nontrivial link-overlap distribution corresponds to space-filling clusters.

A possible way to distinguish between the two scenarios would be to ascertain the true nature of zero temperature excitations. This is precisely what some groups have tried to do, performing a new series of numerical simulations (Houdayer and Martin, 1998; Palassini and Young, 1999, 2000; Krzakala and Martin, 2000). As a result, a new scenario, intermediate between the mean field and the droplet one, has been proposed for the ordered spin glass state (Krzakala and Martin, 2000; Palassini and Young, 2000). Within this picture, there are extensive excitations with *finite* energy (i.e. corresponding to $\theta = 0$), but their surface has fractal dimension $D_s < D$, that is they are not space-filling as would be expected for mean field-like excitations. As a consequence one would get a nontrivial overlap distribution $P(Q)$, but a trivial link-overlap distribution. For this reason, this scenario has been called by Krzakala and Martin the TNT picture (trivial link-overlap, and nontrivial overlap). Marinari and Parisi (2000, 2001) have presented different extrapolations of the numerical data to infinite volume limit and argued that they are compatible with the mean field scenario.

11.2 Conclusions

So, after a long winding road in spin glass land we come to the conclusion.

We have gone through a surprisingly rich and complex amount of results which give a coherent analytic construct for a spin glass field theory. This theory works out for a field ϕ_i^{ab} governed by a Lagrangian with a cubic trace term in w, and an irrelevant dangerous quartic u term. Under mean field it yields the strange and unexpected Parisi order parameter $\langle \phi_i^{ab} \rangle = Q^{ab} \to Q(x)$, infinite in number of components. Working out fluctuation correlations for $\varphi_i^{ab} = \phi_i^{ab} - Q^{ab}$ yields an array of exotic propagators characterized by a spectrum of an infinite number of masses, distributed onto two bands. One band is centred around a large mass $M \sim \tau \sim T_c - T$ with a width of order τ^2. The other band with a small mass $m \sim \tau^2$ sits gapless on top of zero modes.

As long as one works above dimension $D = 8$, mean field gives the correct answer and fluctuations do not affect the result. As one goes below, with $6 < D < 8$, there is a loop contribution to the effective u-coupling that diverges as $\tau^{(D-8)/2}$, a nonproliferating divergence that would usually be ignored. However, if in standard field theories the order parameter is governed by the marginal coupling, in this strange world the irrelevant (dangerous) coupling is also involved. As a result, the characteristic small mass is changed to $m \sim \tau \cdot \tau^{(D-6)/2}$. Thus, as one is approaching $D = 6$, both masses behave like τ (restoring scaling) and leaving a ratio $m/M \sim w^2/z$ (z being the number of neighbours).

Below $D = 6$, the upper critical dimension, fluctuations dominate via a proliferation of logarithmic infrared divergences. The renormalization group works with a stable fixed point, in zero magnetic field, $(w_{\mathrm{R}}^{\star})^2/z \sim \varepsilon$ (see Appendix B: at T_{c} the propagators are trivial and massless, and what is done there for a φ^4 theory is easily repeated for a φ^3 one). Below and away from $D = 6$ the mass ratio becomes an unknown function of ε, with no solid reason to believe that the two bands do not mix. With scaling arguments one can work out predictions for the infrared behaviour of correlation functions which point to a lower critical dimension D_{lc} close to $5/2$. All this under the protection of Ward–Takahashi Identities, ensuring that zero modes obtained at fluctuation level will not, under loop corrections, turn out negative, which would destroy the theory. Clearly, expansions in $\varepsilon = 6 - D$ are not very efficient to predict what is going to happen in $D = 3$. Yet, with some scaling assumptions one can get a variety of analytic predictions.

What is missing so far is the analogue for the spin glass of a nonlinear σ model that would yield results at low temperature and expansions in $D - D_{\mathrm{lc}}$ with $D_{\mathrm{lc}} \sim 5/2$. The same transverse generators that give birth to Ward–Takahashi Identities are the ones that are expected to construct the nonlinear σ-like representation for the field φ_i^{ab}.

Finally, a few words concerning the droplet description of the spin glass. It has always been clear to us that, somehow, the replicated field theory should, besides the mean field-like construct exhibited in the last six chapters, also contain, hidden in the formalism, a formulation that in effect does not break the replica symmetry and will lead to the properties described above that characterize the droplet approach. If only since, after all, the replicated field theory is a representation of the Edwards–Anderson Hamiltonian that the droplet model is supposed to deal with. Steps in identifying that formulation are presently being made (Moore, 2005; De Dominicis, 2005; Temesvari, 2005).

References

Billoire A. and Marinari E. (2002). *Europhys. Lett.*, **60**, 775.
Boettcher S. (2004). *Europhys. Lett.*, **67**, 453.
Boettcher S. (2005). Preprint cond-mat/0508061.
Bokil H., Bray A. J., Drossel B. and Moore M. A. (1999). *Phys. Rev. Lett.*, **82**, 5174, 5177.
Bray A. J. and Moore M. A. (1984). *J. Phys. C*, **17**, L613.

Bray A. J. and Moore M. A. (1986). In *Heidelberg Colloquium on Glassy Dynamics and Optimization*, eds. L. Van Hemmen and I. Morgensten, Heidelberg, Springer-Verlag.
Carter A. C., Bray A. J. and Moore M. A. (2002). *Phys. Rev. Lett.*, **88**, 077201.
Cugliandolo L. F. and Kurchan J. (1994). *J. Phys. A*, **27**, 5749.
De Dominicis C. (2005). *Eur. Phys. J. B*, **48**, 373.
Fisher D. S. and Huse D. A. (1986). *Phys. Rev. Lett.*, **56**, 1601.
Fisher D. S. and Huse D. A. (1987). *J. Phys. A*, **20**, L1005.
Fisher D. S. and Huse D. A. (1988). *Phys. Rev. B*, **38**, 386.
Franz S., Mézard M., Parisi G. and Peliti L. (1999a). *Phys. Rev. Lett.*, **81**, 1758.
Franz S., Mézard M., Parisi G. and Peliti L. (1999b). *J. Stat. Phys.*, **97**, 459.
Gardner E. (1984). *J. Physique*, **45**, 1755.
Houdayer J. and Martin O. C. (1998). *Phys. Rev. Lett.*, **81**, 2554.
Huse D. A. and Fisher D. S. (1987). *J. Phys. A*, **20**, 1997.
Krzakala F. and Martin O. C. (2000). *Phys. Rev. Lett.*, **85**, 3013.
Marinari E. and Parisi G. (2000). *Phys. Rev. B*, **62**, 11677.
Marinari E. and Parisi G. (2001). *Phys. Rev. Lett.*, **86**, 3887.
Marinari E., Parisi G. and Ruiz-Lorenzo J. J. (1998). In *Spin Glasses and Random Fields*, ed. A. P. Young, Singapore, World Scientific.
McMillan W. L. (1984). *J. Phys. C*, **17**, 3179.
Moore M. A. (2005). *J. Phys. A*, **38**, L61.
Moore M. A., Bokil H. and Drossel B. (1998). *Phys. Rev. Lett.*, **81**, 4252.
Newman C. M. and Stein D. L. (1992). *Phys. Rev. B*, **46**, 973.
Newman C. M. and Stein D. L. (1996). *Phys. Rev. Lett.*, **76**, 515.
Newman C. M. and Stein D. L. (1998). *Phys. Rev. E*, **57**, 1356.
Newman C. M. and Stein D. L. (2002). *J. Stat. Phys.*, **106**, 213.
Palassini M. and Young A. P. (1999). *Phys. Rev. Lett.*, **83**, 5126.
Palassini M. and Young A. P. (2000). *Phys. Rev. Lett.*, **85**, 3017.
Ricci-Tersenghi F. and Ritort F. (2000). *J. Phys. A*, **33**, 3727.
Rizzo T. and Crisanti A. (2003). *Phys. Rev. Lett.*, **90**, 137201.
Temesvari T. (2005). *J. Phys. A*, **39**, L783.

Appendix A

Renormalization at one loop: ϕ^4 theory (pure Ising)

Let us consider the *pure* system, i.e. the ϕ^4 model for a pure Ising ferromagnet. The (Helmholtz) free energy $F\{H_i\}$ of the system in the presence of an external source H_i is defined by

$$Z\{H_i\} = \mathrm{e}^{-F\{H_i\}} = \int \prod_i (\mathrm{D}\phi_i) \exp\left\{\mathcal{L}\{\phi_i\} + \sum_i H_i\phi_i\right\} \tag{A.1}$$

where $\mathcal{L}\{\phi_i\}$ can be read off Eq. (2.7) in zero random field. The Gibbs free energy $\Gamma\{M_i\}$ is obtained via the Legendre transform of $F\{H_i\}$:

$$\Gamma\{M_i\} = F\{H_i\} + \textstyle\sum_i H_i M_i, \tag{A.2}$$

$$\frac{\partial F}{\partial H_i} = -M_i, \qquad H_i = \frac{\partial \Gamma}{\partial M_i}, \tag{A.3}$$

with $M_i = \langle \phi_i \rangle$. $\Gamma\{M_i\}$ is the generating function for the N-point vertex functions

$$\Gamma\{M_i\} = \sum_N \frac{1}{N!} \sum_{i_1 \ldots i_N} M_{i_1} \cdots M_{i_N} \Gamma^{(N)}_{(i_1,\ldots,i_N)}. \tag{A.4}$$

To *one loop* (i.e. considering only diagrams with one independent loop) its expression is given by (see (2.48))

$$-\Gamma\{M_i\} = \left.\left\{\mathcal{L}\{\phi_i\} + \frac{1}{2}\mathrm{tr}\ln(-\partial^2\mathcal{L}/\partial\phi_i\,\partial\phi_j) + \cdots\right\}\right|_{\phi_i = M_i}, \tag{A.5}$$

i.e. explicitly for the ϕ^4 model,

$$\begin{aligned} \Gamma\{M_i\} = {} & \frac{1}{2}\int \frac{\mathrm{d}\overset{D}{p}}{(2\pi)^D} M(p)(p^2 + r_0)M(-p) + \frac{u_1}{4!}\sum_i M_i^4 \\ & + \frac{1}{2}\mathrm{tr}\ln\left\{\left[G^0\right]^{-1}_{ij} + \frac{u_1}{2}M_i^2\delta_{ij}\right\}. \end{aligned} \tag{A.6}$$

The N-point vertex functions Γ^N at one loop order can be obtained directly from (A.6) by developing with respect to the M_i.

Let us focus upon $\Gamma^{(4)}$. In Fourier transform we have

$$\frac{1}{4!}\Gamma^{(4)}(p_i) = \frac{1}{4!} \times + \frac{1}{2^4} \times\!\!\bigcirc\!\!\times + \cdots, \quad \text{(A.7)}$$

where $\sum_{i=1}^4 p_i = 0$. Making (A.7) explicit at the transition, we get

$$\Gamma^{(4)}(p_i) = u_1 - \frac{3}{2}u_1^2 \int_0^\wedge \frac{\mathrm{d}q^D}{(2\pi)^D} \frac{1}{q^2(p+q)^2} + \cdots, \quad \text{(A.8)}$$

where $p = p_1 + p_2$. Note that a UV divergence develops at $D = 4$. The renormalization procedure can now be summarized in the following technical steps (Brézin *et al.*, 1976; Zinn-Justin, 1989).

(i) *Define* a renormalized value u_1^R of u_1 for a special, reference value, of the ps, e.g. at the symmetry point $p_i \cdot p_j = (1/3)(4\delta_{ij} - 1)\mu^2$, which will incorporate the UV divergences of $\Gamma^{(4)}$:

$$\Gamma^{(4)}(p_i)|_{SP} \equiv u_1^R \mu^\varepsilon = \times + \frac{3}{2} \times\!\!\bigcirc\!\!\times\Big|_\mu, \quad \text{(A.9)}$$

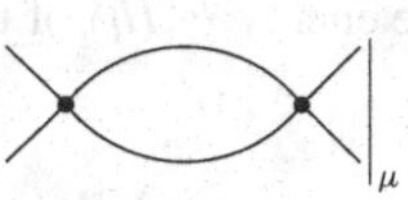

where ε = 4-D and u_1^R is dimensionless. By subtraction we get

$$\Gamma^{(4)}(p_i) = u_1^R\mu^\varepsilon - \frac{3}{2}u_1^2 \left[\times\!\!\bigcirc\!\!\times\Big|_p - \times\!\!\bigcirc\!\!\times\Big|_\mu \right], \quad \text{(A.10)}$$

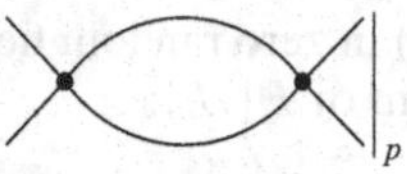

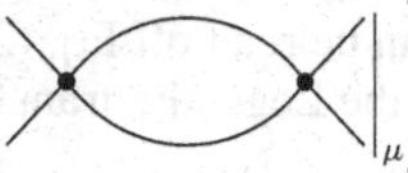

where now the q-integration is convergent in $D = 4$. Since we work at one loop, in the loop term u_1 can be replaced by $u_1^R\mu^\varepsilon$ and we have

$$\begin{aligned} \Gamma^{(4)}(p_i) &= u_1^R\mu^\varepsilon \left\{ 1 - \frac{3}{2} S_D\, u_1^R \frac{\mu^\varepsilon}{\varepsilon} \left[\frac{1}{p^\varepsilon} - \frac{1}{\mu^\varepsilon} \right] + \cdots \right\} \\ &= u_1^R\mu^\varepsilon \left\{ 1 + \frac{3}{2} S_D\, u_1^R \ln(p/\mu) + O\left(\varepsilon \ln^2(p/\mu)\right) \right\}, \end{aligned} \quad \text{(A.11)}$$

where S_D is the surface of the D dimensional sphere. For the explicit normalization condition (A.9), we have

$$u_1^R\mu^\varepsilon = u_1 \left[1 - \frac{3}{2} S_D \frac{u_1}{\varepsilon\mu^\varepsilon} + \cdots \right], \quad \text{(A.12)}$$

which displays the UV divergence that u_1^R is absorbing.

(ii) Strictly speaking, beyond one loop, to take care of all UV divergences, it is also necessary to renormalize the field ϕ_i itself. In this way, for the four-point vertex function, we write

$$Z_\phi^{-2}\, \Gamma_R^{(4)}(p_i; u_1^R, \mu) = \Gamma^{(4)}(p_i; u_1, \wedge), \quad \text{(A.13)}$$

where Z_ϕ is the field renormalization factor. At the one-loop level $Z_\phi = 1$ (and, see below, its associated exponent $\eta = 0$). Expressing that the right hand side of (A.13) is μ independent† at

† Alternatively, we could have also decided to take advantage of the fact that the left hand side of (A.13) is independent of the cut off $\wedge$.

u_1 fixed, one obtains the Callan–Symanzik equation

$$\mu \frac{\mathrm{d}}{\mathrm{d}\mu}\left(Z_\phi^{-2}\,\Gamma_{\mathrm{R}}^{(4)}(p_i; u_1^{\mathrm{R}}, \mu)\right) \equiv \left(\mu \frac{\partial}{\partial \mu} + \beta\left(u_1^{\mathrm{R}}\right)\frac{\partial}{\partial u_1^{\mathrm{R}}} - 2\eta\left(u_1^{\mathrm{R}}\right)\right)\Gamma^{(4)}\left(p_i; u_1^{\mathrm{R}}, \mu\right) = 0 \tag{A.14}$$

with

$$\beta\left(u_1^{\mathrm{R}}\right) \equiv \mu \left.\frac{\partial u_1^{\mathrm{R}}}{\partial \mu}\right|_{u_1}, \tag{A.15}$$

$$\eta\left(u_1^{\mathrm{R}}\right) \equiv \mu \left.\frac{\partial \ln Z_\phi}{\partial \mu}\right|_{u_1}. \tag{A.16}$$

When going away from T_c, the Callan–Symanzik type of equation allows us to derive exponents and scaling relationships. Rewriting (A.12) as

$$u_1 = u_1^{\mathrm{R}} \mu^{\varepsilon}\left[1 + \frac{3}{2} S_D \frac{u_1^{\mathrm{R}}}{\varepsilon} + \cdots\right] \tag{A.17}$$

allows us now to compute $\beta\left(u_1^{\mathrm{R}}\right)$ as of (A.15)[‡], i.e.

$$0 = \varepsilon\, u_1^{\mathrm{R}} + \beta\left(u_1^{\mathrm{R}}\right) + 3\,\frac{u_1^{\mathrm{R}}\beta\left(u_1^{\mathrm{R}}\right)}{\varepsilon} + \frac{3}{2}\left(u_1^{\mathrm{R}}\right)^2, \tag{A.18}$$

that is

$$\beta\left(u_1^{\mathrm{R}}\right) = -\varepsilon\, u_1^{\mathrm{R}} + \frac{3}{2}\left(u_1^{\mathrm{R}}\right)^2 + \cdots \tag{A.19}$$

From the beta function we get the fixed point value $u_1^{\mathrm{R}*}$, such that $\beta(u_1^{\mathrm{R}*}) = 0$, that is

$$u_1^{\mathrm{R}*} = \frac{2}{3}\varepsilon. \tag{A.20}$$

(iii) If we want to know the *flow to or away from* $u_1^{\mathrm{R}*}$, we may ask what happens when we scale the lengths by a factor $1/\rho$ (or the momenta by a factor ρ), then (A.15) becomes

$$\rho \frac{\partial u_1^{\mathrm{R}}}{\partial \rho}(\rho) \equiv \beta\left(u_1^{\mathrm{R}}\right) = -\varepsilon u_1^{\mathrm{R}} + \frac{3}{2}\left(u_1^{\mathrm{R}}\right)^2, \tag{A.21}$$

i.e.

$$u_1^{\mathrm{R}}(\rho) = u_1^{\mathrm{R}*}\,\frac{1}{1 - \frac{u_i^{\mathrm{R}} - u_1^{\mathrm{R}*}}{u_i^{\mathrm{R}}}\,\rho^{\beta'(u^{\mathrm{R}*})}}, \tag{A.22}$$

where $u_1^{\mathrm{R}*}$ is such that $\beta(u_1^{\mathrm{R}*}) = 0$ as in (A.20) and $u_i^{\mathrm{R}} = u_1^{\mathrm{R}}\ (\rho = 1)$ is the *initial value.* For a small deviation from $(u_1^{\mathrm{R}*})$, (A.22) becomes

$$u_1^{\mathrm{R}}(\rho) \simeq u_1^{\mathrm{R}*}\left[1 + \frac{u_i^{\mathrm{R}} - u_1^{\mathrm{R}*}}{u_i^{\mathrm{R}}}\,\rho^{\beta'*}\right], \tag{A.23}$$

where $\beta'(u_1^{\mathrm{R}*}) = \varepsilon$, also called the exponent ω.

Hence the fixed point limit ($\rho \to 0$) is *stable* against small deviations, and independent of the initial value u_i^{R}, i.e. *universal.*

‡ Note that here we have incorporated S_D, the surface of the sphere in D dimensions, in the definition of the couplings.

At the fixed point $u_1^{R^*} = u_1^R\,(\rho = 0)$ we see from (A.14) that $\Gamma^{(4)} \sim p^{\varepsilon}$, as given from simple dimensional analysis (the anomalous part of the ϕ^4 dimension only starts at two loops).

The strong coupling problem (as signalled by the infrared divergences occurring below $D = 4$) is transformed into a weak coupling problem via the transmutation of u_1^R into $u_1^R\,(\rho) \to u_1^{R^*} \sim \varepsilon$, that results from the renormalization property of the theory. From there one can easily compute universal quantities like exponents, amplitude ratios, etc. Our interest here is restricted to finding out that, at the Curie temperature, the solution computed with the coupling fixed point value $u_1^{R^*}$ is *stable*.

References

Brézin E., Le Guillou J. C. and Zinn-Justin J. (1976). In *Phase Transitions and Critical Phenomena,* Vol. 6., eds. C. Domb and M. S. Green, New York, Academic Press.

Zinn-Justin J. (1989). *Quantum Field Theory and Critical Phenomena*, Oxford, Clarendon Press.

Appendix B

Renormalization at one loop: tr ϕ^3 theory (spin glass)

Here we want to obtain the corresponding results to Appendix A for the spin glass. Instead of (A.1), (A.2) and (A.3) we write

$$\overline{Z^n\{H_i^{ab}\}} = \mathrm{e}^{-F\{H_i^{ab}\}} = \int \prod_{\substack{(ab)\\ i}} (\mathrm{D}\phi_i^{ab}) \exp\left\{ \mathcal{L}\{\phi_i^{ab}\} + \sum_i \sum_{(ab)} H_i^{ab}\phi_i^{ab} \right\}, \tag{B.1}$$

with the Lagrangian $\mathcal{L}$ given as in the first line of (8.5), and

$$\Gamma\{Q_i^{ab}\} = F\{H_i^{ab}\} + \textstyle\sum_i H_i^{ab} Q_i^{ab}, \tag{B.2}$$

$$\frac{\partial F}{\partial H_i^{ab}} = -Q_i^{ab}, \qquad H_i^{ab} = \frac{\partial \Gamma}{\partial Q_i^{ab}}, \tag{B.3}$$

with $Q_i^{ab} = \langle \phi_i^{ab} \rangle$.

With the cubic marginal coupling w, the field normalization Z_ϕ appears now at one loop (in the ϕ^4 theory of Appendix A it was showing up at two loops only). To effect the renormalization at one loop in the condensed phase, i.e. below T_c, would be a complicated affair, but working at T_c it drastically simplifies. With $Q^{ab} = 0$, the quadratic form in $\phi_i^{ab} = \varphi_i^{ab}$, i.e. the Hessian, is now diagonal in replica space. Perturbationwise, one works with

bare propagator: $\dfrac{1}{p^2}$

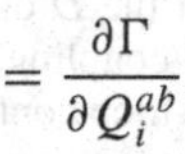

vertex: w

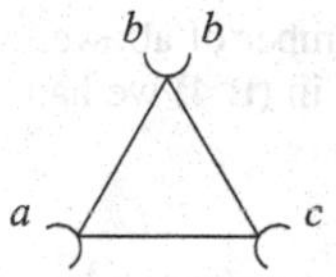

a graphically faithful representation. We have

$$\Gamma^{(2)}_{ab;ab}(p; w, \wedge) = p^2 + \tau_\mathrm{c} - \tag{B.4}$$

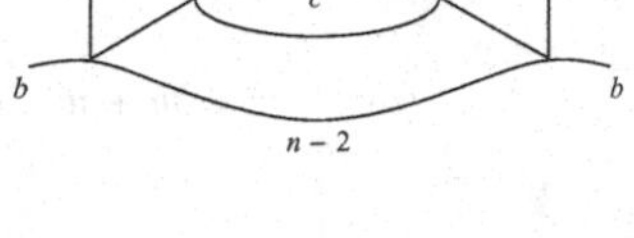

i.e., dropping replica indices,

$$\Gamma^{(2)}(p; w, \wedge) = p^2 + \tau_c - w^2(n-2)\int \frac{d\overset{D}{q}}{(2\pi)^D} \frac{1}{q^2(p+q)^2} \tag{B.5}$$

with, by definition, $\Gamma^{(2)}(0; w, \wedge) = 0$.

Let us now proceed with the renormalization procedure (Harris, Lubensky and Chen, 1976, see also Fischer and Hertz, 1991).

(i) The renormalized $\Gamma_R^{(N)}$ functions are now given by

$$Z_\phi^{-N/2}\,\Gamma_R^{(N)}(p_i; w_R, \mu) = \Gamma^{(N)}(p_i; w, \wedge), \tag{B.6}$$

and the renormalization conditions by

$$\frac{\partial}{\partial p^2}\,\Gamma_R^{(2)}(p; w_R, \mu)\Big|_{p^2=\mu^2} \equiv 1, \tag{B.7}$$

and the analogue of (A.9) is

$$\Gamma_R^{(3)}(p_i; w_R, \mu)\Big|_{SP} \equiv w_R\,\mu^\varepsilon, \tag{B.8}$$

where the momenta p_i are taken at the symmetry point $p_i \cdot p_j = (1/2)(3\delta_{ij} - 1)\mu^2$. From (B.6), (B.7) and using (B.4) one obtains

$$Z_\phi^{-1} = \frac{\partial}{\partial p^2}\Gamma_R^{(2)}(p; w, \wedge)\Big|_{p^2=\mu^2} = 1 + w^2 S_D \frac{(n-2)}{3\varepsilon}\frac{1}{\mu^\varepsilon}, \tag{B.9}$$

where S_D is the volume of the D dimensional sphere, and $\varepsilon = 6 - D$ ($D = 6$ being the upper critical dimension for a ϕ^3 coupling).

Turning now to the one-loop contribution of $\Gamma^{(3)}$, one has

$\Gamma^{(3)}_{ab;bc;ca}(p_i) =$

(B.10)

Here $n - 3$ is the number of allowed values for the index carried by the internal loop (different from a, b, c), just as in (B.4) we had $n - 2$ (the index being different from a, b). Dropping the replica indices one has

$$\Gamma^{(3)}(p_i; w, \wedge) = w + w^3(n-2)\int \frac{d\overset{D}{q}}{(2\pi)^D}\frac{1}{q^2(q+p_1)^2(q+p_1+p_2)^2}. \tag{B.11}$$

Using (B.6), (B.8) and (B.11) taken at the Symmetry Point, we get

$$\begin{aligned} Z_\phi^{-3/2} w_R\,\mu^{\varepsilon/2} &= w + w^3(n-2)\int \frac{d\overset{D}{q}}{(2\pi)^D}\frac{1}{q^2(q^2+\mu^2)^2} \\ &= w + w^3 S_D \frac{(n-2)}{\varepsilon}\frac{1}{\mu^\varepsilon}. \end{aligned} \tag{B.12}$$

To lowest order $w^3 \sim w_{\mathrm{R}}^3\, \mu^{3\varepsilon/2}$ and hence

$$w = \left(Z_\phi^{-3/2} w_{\mathrm{R}} - w_{\mathrm{R}}^3 S_D \frac{(n-2)}{\varepsilon} \right) \mu^{\varepsilon/2}, \tag{B.13}$$

and likewise, from (B.9),

$$Z_\phi = 1 - w_{\mathrm{R}}^3 S_D \frac{(n-2)}{3\,\varepsilon}. \tag{B.14}$$

(ii) We can now turn to the Callan–Symanzik equation resulting from the r.h.s. of Eq. (B.6) being μ independent (at fixed w), that is, as in (A.14),

$$\begin{aligned} &\mu \frac{\mathrm{d}}{\mathrm{d}\mu} \left(Z_\phi^{-2}\, \Gamma_{\mathrm{R}}^{(N)}(p_i; w_{\mathrm{R}}, \mu) \right) \\ &\equiv \left(\mu \frac{\partial}{\partial \mu} + \beta(w_{\mathrm{R}}) \frac{\partial}{\partial w_{\mathrm{R}}} - \frac{N}{2} \eta(w_{\mathrm{R}}) \right) \Gamma_{\mathrm{R}}^{(N)}(p_i; w_{\mathrm{R}}, \mu) = 0, \end{aligned} \tag{B.15}$$

with

$$\beta(w_{\mathrm{R}}) \equiv \mu \left. \frac{\partial w_{\mathrm{R}}}{\partial \mu} \right|_w , \tag{B.16}$$

$$\eta(w_{\mathrm{R}}) \equiv \mu \left. \frac{\partial \ln Z_\phi}{\partial \mu} \right|_w , \tag{B.17}$$

yielding, with (B.9), (B.13) and (B.14),

$$\beta(w_{\mathrm{R}}) = -w_{\mathrm{R}} \left(\frac{\varepsilon}{2} - \frac{3}{2}\eta \right) - w_{\mathrm{R}}^3 S_D (n-2), \tag{B.18}$$

$$\eta(w_{\mathrm{R}}) = w_{\mathrm{R}}^2 S_D \frac{(n-2)}{3}. \tag{B.19}$$

(iii) The fixed point given by $\beta(w_{\mathrm{R}}^*) = 0$ comes out as

$$(w_{\mathrm{R}}^*)^2 S_D = \frac{\varepsilon}{2-n}, \tag{B.20}$$

and

$$\eta = -\frac{\varepsilon}{3}\,. \tag{B.21}$$

The fixed point is *stable*, with $\omega = +\varepsilon$. Note that at the fixed point Eq. (B.15) yields

$$\Gamma_{\mathrm{R}}^{(2)}(p; \mu) \sim p^2 \left(\frac{p}{\mu} \right)^{-\eta}, \tag{B.22}$$

identifying η as the anomaly.

In both the appendices, had we kept the factors $1/z$ that count the number of loops, we would have got for the fixed points, respectively,

$$u_1^{\mathrm{R}*} \frac{S_D}{z} = \frac{2\varepsilon}{3} \qquad \text{with } \varepsilon = 4 - D, \tag{B.23}$$

$$(w_{\mathrm{R}}^*)^2 \frac{S_D}{z} = \frac{\varepsilon}{2-n} \qquad \text{with } \varepsilon = 6 - D. \tag{B.24}$$

References

Fischer K. H. and Hertz J. A. (1991). *Spin Glasses*, Cambridge, Cambridge University Press.

Harris A., Lubensky T. and Chen J. (1976). *Phys. Rev. Lett.*, **36**, 415.

Index